AF517839

ENCYCLOPAEDIA OF ENGINEERING

ENCYCLOPAEDIA OF ENGINEERING

VOL. 4
ELECTRICAL ENGINEERING

By
A.M. Kharmawphlang
A. Khongphai

AKANSHA PUBLISHING HOUSE
NEW DELHI

AKANSHA PUBLISHING HOUSE
4649-B/21, Ansari Road
Darya Ganj, New Delhi-110 002
Email: ektabooks@yahoo.com
Ph.: 23263193/9811582579
Fax: 011-23263193

Encyclopaedia of Engineering

First Published 2006

ISBN 81-8370-048-9 (Set)

[The responsibility for the facts stated, conclusions reached etc. is entirely that of the authors. The publisher is not responsible for them, whatsoever.]

PRINTED IN INDIA

Published by M.P. Misra for Akansha Publishing House, New Delhi and Printed at Aryan Enterprise, Delhi.

PREFACE

Engineering is the application of science to the needs of humanity. This is accomplished through the application of knowledge, mathematics, and practical experience to the design of useful objects or processes.

Engineering involves the knowledge of mathematical and natural sciences (biological and physical) gained by study, experience, and practice, applied with judgement and creativity to develop ways to utilize the materials and forces of nature for the benefit of mankind.

Encyclopaedia of engineering contains engineering terms with clear and concise definitions. The encyclopaedia is highly illustrated and written in plain, non-technical language and extensive cross-referencing makes concepts more easily understood. Timings and Twigg focus on a wide range of engineering disciplines — engineering science, electrical and electronic engineering, workshop practices and mechanical engineering — and provide definitions that are relevant to all.

This vital reference offers a wealth of essential information in a convenient, quick-find format. Whether you're a professional, a student, a writer, or a general reader with an interest in science, there is no better or more authoritative way to stay up-to-speed with the current language of the various branches of engineering or gain an understanding of their key ideas and concepts.

Encyclopaedia of Engineering is written specifically for further education students by experienced authors who have taught and written extensively at this level.

A

A. Abbreviation for "ampere" a unit of electrical current.

AAR (Automatic Alternative Routing). A system for providing continued telecommunications service in the event that a primary transmission route fails catastrophically (e.g., when a FOTS cable is cut).

A-Band Cellular. In the U.S. cellular duopoly, the alternative carrier to the regional Bell operating company's cellular subsidiary.

Abrasive. Any of a number of hard materials, such as aluminum oxide, silicon carbide, and diamond, that are powdered and carefully graded according to particle size, and used to shape and/or finish optical elements, including the endfaces of optical fibers and connectors. *Note:* For finishing the endfaces of optical fiber connectors, abrasive particles are adhered to a substrate of plastic film, in a fashion after that of sandpaper. The film is in turn supported by a hard, flat plate. The connector is supported by a fixture that holds it securely in the proper position for finishing. The grinding motion may be performed manually or by a machine.

Absolute Gain. 1. Of an antenna, for a given direction and polarization, the ratio of (a) the power that would be required at the input of an ideal isotropic radiator to (b) the power actually supplied to the given antenna, to produce the same radiation intensity in the far-field region. *Note 1:* If no direction is given, the absolute gain of an antenna corresponds to the direction of maximum effective radiated power. *Note 2:* Absolute gain is usually expressed in dB. *Synonym* isotropic gain. 2. Of a device, the ratio of (a) the signal level at the output of the device to (b) that of its input under a specified set of operating conditions. *Note 1:* Examples of absolute gain are no-load gain, full-load gain, and small-signal gain. *Note 2:* Absolute gain is usually expressed in dB.

Absorber. A material that readily absorbs photons to generate charge carriers (free electrons or holes).

Absorbers. Dark-colored objects that soak up heat in solar collectors.

Absorption. 1. Loss or dissipation of energy as it travels through a medium, Example: radio waves lose some of their energy as they travel through the atmosphere.

2. In the transmission of electrical, electromagnetic, or acoustic signals, the conversion of the transmitted energy into another form, usually thermal. *Note 1:* Absorption is one cause of signal attenuation. *Note 2:* The conversion takes place as a result of interaction between the incident energy and the material medium, at the molecular or atomic level.

Absorption Coefficient. 1. The factor by which photons are absorbed as they travel a unit distance through a material. 2. A measure of the attenuation caused by absorption of energy that results from its passage through a medium. *Note 1:* Absorption coefficients are usually expressed in units of reciprocal distance. *Note 2:* The sum of the absorption coefficient and the scattering coefficient is the attenuation coefficient.

Absorption Modulation. Amplitude modulation of the output of a radio transmitter by means of a variable-impedance circuit that is caused to absorb carrier power in accordance with the modulating wave.

A/B Switch. Allows a user to change their wireless phone from one carrier to another through the keyboard (older phones actually had a physical switch to do the same thing).

Abbreviated Dialing. A feature on wireless phones where you enter just one or two digits from the keypad and then initiate the call. The phone searches its speed dial directory and associates the entire number with the two-digit speed dial position you've entered.

AC. Abbreviation for "Alternating Current"

Acceptance Angle. In fiber optics, half the vertex angle of that cone within which optical power may be coupled into bound modes of an optical fiber. *Note 1:* The axis of the cone is collinear with the fiber axis, the vertex of the cone is on the fiber end-face, and the base of the cone faces the optical power source. *Note 2:* The acceptance angle is measured with respect to the fiber axis. *Note 3:* Rays entering an optical fiber at angles greater than the acceptance angle are coupled into unbound modes.

Acceptance Cone. In fiber optics, the cone within which optical power may be coupled into the bound modes of an optical fiber. *Note:* The acceptance cone is derived by rotating the acceptance angle about the fiber axis.

Acceptor. A dopant material, such as boron, which has fewer outer shell electrons than required in an otherwise balanced crystal structure, providing a hole, which can accept a free electron.

Acceptor Atoms. Trivalent atoms that accept free electrons from pentavalent atoms.

Access Charge. The fixed part of the monthly fee that carriers charge to subscribers. Sometimes this charge includes a certain amount of minutes of usage.

Access Denial. 1. Access failure caused by the issuing of a system blocking signal by a communications system that does not have a call-originator camp-on feature. 2. Access failure caused by exceeding the maximum access time and nominal system access time fraction during an access attempt. *Synonym* system blocking.

Access Failure. In a communications system, an unsuccessful access that results in termination of an access attempt in any manner other than initiation of user information transfer between the intended source and destination (sink) within the specified maximum access time. *Note:* Access failure can be the result of access denial, access outage, user blocking, or incorrect access.

Access Fee. A special fee that local phone companies are allowed to charge customers for the right to connect with the local phone network. Cellular subscribers contribute to access fees and pay a federal 3 percent telephone excise tax.

Access Point. 1. A point where connections may be made for testing or using particular communications circuits. 2. In telephony, a junction point in outside plant consisting of a semipermanent splice at a junction between a branch feeder cable and distribution cables.

Access Time. 1. In a telecommunication system, the elapsed time between the start of an access attempt and successful access. *Note:* Access time values are measured only on access attempts that result in successful access. 2. In a computer, the time interval between the instant at which an instruction control unit initiates a call for data and the instant at which delivery of the data is completed. 3. The time interval between the instant at which storage of data is requested and the instant at which storage is started. 4. In magnetic disk devices, the time for the access arm to reach the desired track and the delay for the rotation of the disk to bring the required sector under the read-write mechanism.

AC Coupling. Circuit that passes an AC signal while blocking a DC voltage.

AC/DC. Equipment that will operate on either an AC or DC power source.

AC-DC Ringing. Telephone ringing that makes use of both ac and dc voltages and currents. *Note:* An alternating current may be used to operate a ringer and direct current to aid the relay action that stops the ringing when the called telephone is answered.

AC Generator. Device used to transform mechanical energy into AC electrical power.

AC Load Line. A graph representing all possible combinations of AC output voltage and current for an amplifier.

Acoustic Coupler. 1. An interface device for coupling electrical signals by acoustical meansCusually into and out of a telephone instrument. 2. A terminal device used to link data terminals and radio sets with the telephone network. *Note:* The link is achieved through acoustic (sound) signals rather than through direct electrical connection.

Acoustic Echo Canceller (AEC). A full-duplex audio technology used to eliminate acoustically coupled return echoes within a teleconference room. AECs model a room's acoustics (primarily reverberation time and amplitude) and apply the inverse to the outgoing signals to remove artifacts of the incoming signal, enhancing a room's apparent echo return loss. Technologies used may include an AFIRF, center clipper, noise reducer, TEC, AGC and mild echo suppression. See also *tail time, echo suppressor* and *echo return loss enhancement*.

Acoustic Wave. A longitudinal wave that (a) consists of a sequence of pressure pulses or elastic displacements of the material, whether gas, liquid, or solid, in which the wave propagates, (b) in gases, consists of a sequence of compressions (dense gas) and rarefactions (less dense gas) that travel through the gas, (c) in liquids, consists of a sequence of combined elastic deformation and compression waves that travel though the liquid, and (d) in solids, consists of a sequence of elastic compression and expansion waves that travel though the solid. *Note 1:* The speed of an acoustic wave in a.material medium is determined by the temperature, pressure, and elastic properties of the medium. In air, acoustic waves propagate at 332 m/s (1087 ft/s) at 0EC, at sea level. In air, sound-wave speed increases approximately 0.6 m/s (2 ft/s) for each kelvin above 0EC. *Note 2:* Acoustic waves audible to the normal human ear are termed *sound waves*.

Acousto-optic Effect. A variation of the refractive index of a material caused by interaction with acoustic energy in the form of a wave or pulse. *Note:* The acousto-optic effect is used in devices that modulate or deflect light.

Acousto-optics. The discipline devoted to the interactions between acoustic waves and light waves in a material medium. *Note:* Acoustic waves can be made to modulate, deflect, and/or focus light waves by causing a variation in the refractive index of the medium.

AC Power Supply. Power supply that delivers an AC voltage.

Acquisition. 1. In satellite communications, the process of locking tracking equipment on a signal from a communications satellite. 2. The process of achieving synchronization. 3. In servo systems, the process of entering the boundary conditions that will allow the loop to capture the signal and achieve lock-on.

ACR. *Absolute Category Rating*, an ITU-T P.800 subjective listening test performed by HEI.

ACSB. *Amplitude Companded Single sideBand*, an LMR technology providing voice channels with 5 KHz spacing at 218-220 MHz in the USA.

Activation Fee. A charge imposed by carriers for the processing and activation of wireless phone service.

Active Component. A component that changes the amplitude of a signal between input and output.

Active Device. A device that requires a source of energy for its operation and has an output that is a function of present and past input signals. *Note:* Examples of active devices include controlled power supplies, transistors, LEDs, amplifiers, and transmitters.

Active Filter. A filter that uses an amplifier in addition to reactive components to pass or reject selected frequencies.

Active Laser Medium. Within a laser, the material that emits coherent radiation or exhibits gain as the result of electronic or molecular transitions to a lower energy state or states, from a higher energy state or states to which it had been previously stimulated. *Note:* Examples of active laser media include certain crystals, gases, glasses, liquids, and semiconductors. *Synonym* laser medium.

Active Region. The region of BJT operation between saturation and cutoff used for linear amplification.

Active Sensor. 1. A detection device that requires input energy from a source other than that which is being sensed. *Note:* An example of an active sensor is a photoconductive cell. 2. In surveillance, a detection device that emits energy capable of being detected by itself. *Note:* An example of an active sensor is a measuring instrument that generates a signal, transmits it to a target, and receives a reflected signal from the target. Information concerning the target is obtained by comparison of the received signal with the transmitted signal. 3. A measuring instrument in the Earth exploration-satellite service or in the space research service by means of which information is obtained by transmission and reception of radio waves.

Active Solar Heater. A solar water or space-heating system that moves heated air or water using pumps or fans.

ActiveX. A set of Microsoft plug-in program software devices or program modules. When embedded in web pages, often used to add multimedia functionality to the web browser.

AC Voltage. A voltage in which the polarity alternates.

ADC. Abbreviation for "analog to digital converter"

Adaptive Array Antennas. A type of advanced smart antenna technology that continually monitors a received signal and dynamically adapts signal patterns to optimize wireless system performance. The arrays use signal processing algorithms to adapt to user movement, changes in the radio-frequency environment and multipath and co channel interference.

Adaptive Power Control. Technique employed by wireless infrastructure systems that lowers the power of a signal in a cell site whenever the site detects that the user's phone is close to the source of the signal. This saves power in the phone, too, but the cell site tells the phone to lower its power, thus saving battery life.

Adaptive Communications. Any communications system, or portion thereof, that automatically uses feedback information obtained from the system itself or from the signals carried by the system to modify dynamically one or more of the system operational parameters to improve system performance or to resist degradation. *Note:* The modification of a system parameter may be discrete, as in hard-switched diversity reception, or may be continuous, as in a predetection combining algorithm.

Adaptive Predictive Coding (APC). Narrowband analog-to-digital conversion that uses a one-level or multilevel sampling system in which the value of the signal at each sampling instant is predicted according to a linear function of the past values of the quantized signals. *Note:* APC is related to linear predictive coding (LPC) in that both use adaptive predictors. However, APC uses fewer prediction coefficients, thus requiring a higher sampling rate than LPC.

Adaptive Radio. A radio that (a) monitors its own performance, (b) monitors the path quality through sounding or polling, (c) varies operating characteristics, such as frequency, power, or data rate, and (d) uses closed-loop action to optimize its performance by automatically selecting frequencies or channels.

Adder. 1. A device whose output data are a representation of the sum of the numbers represented by its input data. *Note:* An adder may be serial or parallel, digital or analog. 2. A device whose output data are a representation of the sum of the quantities represented by its input data. *Note:* An adder can add things other than representations of numbers. It can add voltages, *etc.* Analog adders are not limited to summing representations of numbers. An adder may operate on digital or analog data.

Adjacent Channel Interference. 1. Signal impairment to one frequency due to presence of another signal on a nearby frequency.

2. Extraneous power from a signal in an adjacent channel. *Note 1:* Adjacent channel interference may be caused by inadequate filtering, such as incomplete filtering of unwanted modulation products in frequency modulation (FM) systems, improper tuning, or poor frequency control, in either the reference channel or the interfering channel, or both. *Note 2:* Adjacent-channel interference is distinguished from crosstalk.

Admittance. (symbol "Y") Measure of how easily AC will flow through a circuit. Admittance is the reciprocal of impedance and is measured in siemens.

ADPCM. *Adaptive Differential Pulse Code Modulation*. Differential pulse code modulation that also uses adaptive quantizing; an audio coding algorithm providing a modest degree of compression together with good quality.

ADSL. *Asymmetrical Digital Subscriber Line*, a modest speed, limited distance, local access solution using a single regular telephone pair. Although an 8 Mbps maximum is theoretically possible, with a maximum return path of 800 Kbps, a 2004 survey showed that in North America, the average bit rate from CO to the subscriber is only about 500 Kbps. The path length cannot exceed 5.5 Km or 3.4 miles. Supports simultaneous transmission of data and voice over the same pair.

Aerial Insert. In a direct-buried or underground cable run, a cable rise to a point above ground, followed by an overhead run, *e.g.*, on poles, followed by a drop back into the ground. *Note:* An aerial insert is used in places where it is not possible or practical to remain underground, such as might be encountered in crossing a deep ditch, canal, river, or subway line.

Aeronautical Fixed Service. A radiocommunication service between specified fixed points provided primarily for the safety of air navigation and for the regular, efficient and economical operation of air transport.

AF. Abbreviation for "audio frequency".

AFC. Abbreviation for "automatic frequency control".

Affiliate. Companies that assist larger carriers with building out a nationwide network; the affiliate may use the primary carrier's brand name, network operations, customer service or other facilities.

AFIRF. *Adaptive Finite Impulse Response Filter*, a time domain filter, important in acoustic echo cancellers (AEC) and in removal of some types of noise from sound signals.

AGC. Abbreviation for "automatic gain control".

AGM. Absorbed Glass Mat, a newer type of battery construction that uses saturated absorbant glass mats rather than gelled or liquid electrolyte. Somewhat more expensive than flooded (liquid), but offers very good reliability.

AGT. AudioGraphics Terminal.

AIN (Advanced Intelligent Network). Introduced by AT&T Network Systems in 1991. Enables service providers to define, test and introduce new multimedia messaging, PCS and cell routing.

Airborne Radio Relay. 1. Airborne equipment used to relay radio transmission from selected originating transmitters. 2. A technique employing aircraft fitted with radio relay stations for the purpose of increasing the range, flexibility, or physical security of communications systems.

Aircraft Emergency Frequency. An international aeronautical emergency frequency, such as 121.5 MHz (civil) and 243.0 MHz (military), for aircraft stations and stations concerned with safety and regulation of flight along national or international civil air routes and maritime mobile service stations authorized to communicate for safety purposes.

Air Interface. The standard operating system of a wireless network; technologies include AMPS, TDMA, CDMA and GSM.

Air Mass. The ratio of the mass of atmosphere in the actual observer-sun path to the mass that would exist if the observer was at sea level, at standard barometric pressure, and the sun was directly overhead. Note—(sometimes called air mass ratio). Air mass varies with the zenith angle of the sun and the local barometric pressure, which changes with altitude. For sun zenith angle, Z, of 62 degrees or less and local atmospheric pressure, P, where Po is standard atmospheric pressure, AM approximately equal sec Z(P/Po). [ASTM E 772] Alternatively, the path length of light through the atmosphere is described in terms of an equivalent relative air mass. AM0 corresponds to the solar spectrum in outer space; at the equator, the average spectrum is AM1, and the reference spectrum for STC was defined to be AM1.5 (average spectrum at 45-degrees latitude).

Air Mass 1.5 (AM1.5) Standard Reference Spectrum. The solar spectral irradiance distribution (diffuse and direct) incident at sea level on a sun-facing 37-degree tilted surface. The atmospheric conditions for AM1.5 are: precipitable water vapor, 14.2 mm; total ozone, 3.4 mm; turbidity (base e, lambda=0.5 mm), 0.27. [ASTM E 892, Table 2]

Airtime. Actual time spent using a wireless phone.

A-key. A secret number issued to a cellular phone that is used in conjunction with a subscriber's shared secret data information for authentication

A-law Algorithm. A standard compression algorithm, used in digital communications systems of the European digital hierarchy, to optimize, *i.e.,* modify, the dynamic range of an analog signal for digitizing. *Note 1:* The wide dynamic range of speech does not lend itself well to efficient linear digital encoding. A-law encoding effectively reduces the dynamic range of the signal, thereby increasing the coding efficiency and resulting in a signal-to-distortion ratio that is superior to that obtained by linear encoding for a given number of bits.

Algorithm. In audio, video and data coding, the step-by-step mathematical process or procedure (often including repetition) providing suitable encoding, compression and/or encryption for a specific application. When used for either lossless or lossy compression, a significant reduction in the number of bits required for transmission takes place.

Alias, Aliasing. A false signal produced by the analog-to-digital sampling process, often caused by failure to observe the *Nyquist criterion,* creating subjectively disturbing artifacts.

Alkaline Cell. A primary cell that delivers more current than a carbon-zinc cell. Also known as an "alkaline manganese cell".

Alligator Clip. Spring clip on the end of a test lead used to make a temporary connection.

All-silica Fiber. An optical fiber composed of a silica-based core and cladding. *Note:* The presence of a protective polymer overcoat does not disqualify a fiber as an all-silica fiber, nor does the presence of a tight buffer. *Synonym* all-glass fiber.

Alpha. Ratio of collector current to emitter current in a bipolar junction transistor (BJT). Greek letter alpha "a" is the symbol used.

Alphanumeric. A display, message or readout that contains both letters and numbers. Synonymous with text paging or messaging.

Alternating Current. 1. An electric current that rises to a maximum in one direction, falls back to zero and then rises to a maximum in the opposite direction and then repeats. 2. Electric current in which the direction of flow is reversed at frequent intervals—usually 100 or 120 times per second (50 or 60 cycles per second or 50//60 Hz).

Alternator. Name for an AC generator.

AM. *Amplitude Modulation,* as in the AM broadcast band.

Ambient Noise Level. The level of acoustic noise existing at a given location, such as in a room, in a compartment, or at a place out of doors. *Note 1:* Ambient noise level is measured with a sound level meter. *Note 2:* Ambient noise level is usually measured in dB above a reference pressure level of 0.00002 Pa, *i.e.,* 20 ìPa (micropascals) in SI units. A pascal is a newton per square meter. *Note 3:* In the centimeter-gram-second system of units, the reference level for measuring ambient noise level is 0.0002 dyn/cm^2. *Synonym* room noise level.

AMI. *Alternate Mark Inversion,* a 1's density control mechanism for T1 and E1 systems. See also B8ZS and HDB3.

Ammeter. A meter used to measure current.

Amorphous Semiconductor. A non-crystalline semiconductor material that has no long-range order. Easier and cheaper to make than crystalline, but less efficient and slowly degrades over time. Also called thin film.

Ampere (A) or amp. The unit for the electric current; the flow of electrons. One amp is 1 coulomb passing in one second. One amp is produced by an electric force of 1 volt acting across a resistance of 1 ohm.

Ampere-hour (Ah). Quantity of electricity or measure of charge. (1 Ah = 3600 C [Coulomb])

Amplifier. A circuit that increases the voltage, current, or power of a signal.

Amplitude. Magnitude or size of a signal voltage or current.

Amplitude Distortion. Distortion occurring in a system, subsystem, or device when the output amplitude is not a linear function of the input amplitude under specified conditions. *Note:* Amplitude distortion is measured with the system operating under steady-state conditions with a sinusoidal input signal. When other frequencies are present, the term *"amplitude"* refers to that of the fundamental only.

Amplitude Modulation (AM). Modulation in which the amplitude of a carrier wave is varied in accordance with some characteristic of the modulating signal. *Note:* Amplitude modulation implies the modulation of a coherent carrier wave by mixing it in a nonlinear device with the modulating signal to produce discrete upper and lower sidebands, which are the sum and difference frequencies of the carrier and signal. The envelope of the resultant modulated wave is an analog of the modulating signal. The instantaneous value of the resultant modulated wave is the vector sum of the corresponding instantaneous values of the carrier wave, upper sideband, and lower sideband. Recovery of the modulating signal may be by direct detection or by heterodyning.

AMPS (Advanced Mobile Phone Service). The analog cellular standard.

AMSS. *Aeronautical Mobile Satellite Service.*

AMTA (American Mobile Telecommunications Association). Washington, D.C.-based trade group representing specialized mobile radio operators.

Analog. 1. Information represented as continuously varying voltage or current rather than in discrete levels as opposed to digital data varying between two discrete levels.

2. The original form of cellular service, launched in October, 1983 in the U.S. (and earlier elsewhere). This service uses a waveform transmission instead of the zeros and ones that a digital system uses. It is more prone to interference, static, eavesdropping and cloning than digital systems, but is still deployed in many parts of the world where the advanced technology (and higher cost) of digital systems is not deemed necessary.

Analog Decoding. The portion of the digital-to-analog conversion process that generates an analog signal value from the digital signal that resulted from analog encoding. *Note:* Further action is required to integrate these samples to obtain a continuous approximation of the original signal, because analog decoding does not smooth the signal.

Analog Encoding. The portion of the analog-to-digital conversion process that samples an analog signal and creates a digital signal that represents the value of the sample. *Note:* Multiple samples are needed to digitize a waveform over a time interval.

Analog Signal. 1. A signal that has a continuous nature rather than a pulsed or discrete nature. *Note:* Electrical or physical analogies, such as continuously varying voltages, frequencies, or phases, may be used as analog signals. 2. A nominally continuous electrical signal that varies in some direct correlation with another signal impressed on a transducer. *Note:* For example, an analog signal may vary in frequency, phase, or amplitude in response to changes in physical phenomena, such as sound, light, heat, position, or pressure.

Angle Modulation. Modulation in which the phase or frequency of a sinusoidal carrier is varied. *Note:* Phase and frequency modulation are particular forms of angle modulation.

Angle of Incidence. Angle between the normal to a surface and the direction of incident radiation; applies to the aperture plane of a solar collector. Most modern solar panels have only minor reductions in power output within plus/minus 15 degrees.

Angular Misalignment Loss. Power loss caused by the deviation from optimum angular alignment of the axes of source to waveguide, waveguide to waveguide, or waveguide to detector. *Note 1:* The waveguide may be dielectric (an optical fiber) or metallic. *Note 2:* Angular misalignment loss does not include lateral offset loss and longitudinal offset loss.

ANI. Automatic Number Identification, in telephony.

Anisotropic. Pertaining to a material whose electrical or optical properties vary with (a) the direction of propagation of a traveling wave or with (b) different polarizations of a traveling wave. *Note 1:* Anisotropy is exhibited by non-cubic crystals, which have different refractive indices for lightwaves propagating in different directions or with different polarizations. *Note 2:* Anisotropy may be induced in certain materials under mechanical strain.

Annual Solar Savings. The annual solar savings of a solar building is the energy savings attributable to a solar feature relative to the energy requirements of a nonsolar building.

Anode. The positive electrode or terminal of a device. The "P" material of a diode.

ANSI. American National Standards Institute, a globally recognized organization issuing IT and computing technical standards.

Antenna. A metallic rod that typically extends from a wireless phone or cell site from which the electrical signal that is transmitted emanates from. Cell sites might have different antennas for transmitting and receiving. Wireless phones might have either small, fixed antennas (called "stubbies") or retractable antennas. Newer phone models have "intennas," which are antennas that are not visible to the user because they are housed completely inside the phone unit.

Antenna Effective Area. The functionally equivalent area from which an antenna directed toward the source of the received signal gathers or absorbs the energy of an incident electromagnetic wave. *Note 1:* Antenna effective area is usually expressed in square meters. *Note 2:* In the case of parabolic and horn-parabolic antennas, the antenna effective area is about 0.35 to 0.55 of the geometric area of the antenna aperture.

Antenna Efficiency. The ratio of the total radiated power to the total input power. *Note:* The total radiated power is the total input power less antenna dissipative losses.

Antenna Electrical Beam Tilt. The shaping of the radiation pattern in the vertical plane of a transmitting antenna by electrical means so that maximum radiation occurs at an angle below the horizontal plane.

Antenna Gain. The ratio of the power required at the input of a loss-free reference antenna to the power supplied to the input of the given antenna to produce, in a given direction, the same field strength at the same distance. *Note 1:* Antenna gain is usually expressed in dB. *Note 2:* Unless otherwise specified, the gain refers to the direction of maximum radiation. The gain may be considered for a specified polarization. Depending on the choice of the reference antenna, a distinction is made between:

Cabsolute or isotropic gain (G_i), when the reference antenna is an isotropic antenna isolated in space;

Cgain relative to a half-wave dipole (G_d) when the reference antenna is a half-wave dipole isolated in space and with an equatorial plane that contains the given direction;

Cgain relative to a short vertical antenna (G_r), when the reference antenna is a linear conductor, much shorter than one quarter of the wavelength, normal to the surface of a perfectly conducting plane which contains the given direction. *Synonyms* gain of an antenna, power gain of an antenna.

Antenna Noise Temperature. The temperature of a hypothetical resistor at the input of an ideal noise-free receiver that would generate the same output noise power per unit bandwidth as that at the antenna output at a specified frequency. *Note 1:* The antenna noise temperature depends on antenna coupling to all noise sources in its environment as well as on noise generated within the antenna. *Note 2:* The antenna noise temperature is a measure of noise whose value is equal to the actual temperature of a passive device.

Antenna, Transmitting. A device that converts an electrical wave into an electromagnetic wave that radiates away from the antenna.

Antenna, Receiving. A device that converts a radiated electromagnetic wave into an electrical wave.

Anthropogenic. Referring to alterations in the environment due to the presence or activities of humans.

Antireflection Coating. A thin, dielectric or metallic film, or several such films, applied to an optical surface to reduce its reflectance and thereby increase its transmittance. *Note:* For minimum reflection of a normal incident wave of a single wavelength, the antireflection coating may consist of a single layer and must have (a) a refractive index equal to the square root of the refractive indices of the materials bounding the coating, and (b) a thickness equal to one-quarter the wavelength in question (*i.e.*, the wavelength within the material of which the coating consists). For minimum reflection of multiple wavelengths, additional layers must be added.

API. *Application Programming Interface*, a set of software routines useful for interfacing between high level systems (such as users) and low-level systems (such as operating systems or system level tasks). Often the design basis for a man-machine interface.

Apparent Power. In alternating-current power transmission and distribution, the product of the rms voltage and amperage. *Note 1:* When the applied voltage and the current are in phase with one another, the apparent power is equal to the effective power, *i.e.*, the real power delivered to or consumed by the load. If the current lags or leads the applied voltage, the apparent power is greater than the effective power. *Note 2:* Only effective power, *i.e.*, the real power delivered to or consumed by the load, is expressed in watts. Apparent power is properly expressed only in volt-amperes, never watts. *See diagram under* effective power.

Arc. Discharge of electricity through a gas such as lightning discharging through the atmosphere.

Area Loss. When optical fibers are joined by a splice or a pair of mated connectors, a power loss that is caused by any mismatch in size or shape of the cross section of the cores of the mating fibers. *Note 1:* Any of the above conditions may allow light from the core of the "transmitting" fiber to enter the cladding of the "receiving" fiber, where it is quickly lost. *Note 2:* Area loss may be dependent on the direction of propagation. For example, in coupling a signal from an optical fiber having a smaller core to an otherwise identical one having a larger core, there will be no area loss, but in the opposite direction, there will be area loss.

ARIB (Association of Radio Industries and Businesses). The Japanese standards-setting organization.

Armature. The rotating or moving component of a magnetic circuit.

Armor. Of a communications cable, a component intended to protect the critical internal components, *e.g.*, buffer tubes or fibers, or electrical conductors, from damage from external mechanical attack, *e.g.*, rodent attack or abrasion. *Note:* Armor usually takes the form of a steel or aluminum tape wrapped about an inner jacket that covers the critical internal components. An outer jacket usually covers the armor.

Armstrong Oscillator. An oscillator that uses an isolation transformer to achieve positive feedback from output to input.

ARPU. Average revenue per user. One indicator of wireless carrier progress, generally divulged quarterly.

Array. Any number of photovoltaic modules connected together to provide a single electrical output. Arrays are often designed to produce significant amounts of electricity.

ARS. Automatic Route Selection, an important component of an automatic least-cost routing system.

Artifact. Spurious effects or imperfections introduced into a signal as a result of digital signal processing.

ASIC. Application-Specific Integrated Circuit

ASCII. American Standard Code for Information Interchange. A standard set of alphanumeric and control codes representing unformatted text, which can be represented by a single byte, developed by the USA Standards Institute. In telecommunications, ASCII is often the coding used for the control of telecommunications equipment. HTML tags and URLs are examples of use of ASCII.

ASP. *Application Service Provider*, a company that provides third-party services and software over the Internet. *Active Server Page*, a type of web page that is an interface with one type of program that runs on a web server, often providing custom content for users. Used commonly for e-commerce.

Assigned Frequency. 1. The center of the assigned frequency band assigned to a station. 2. The frequency of the center of the radiated bandwidth. *Note:* The frequency of the rf carrier, whether suppressed or radiated, is usually given in parentheses following the assigned frequency, and is the frequency appearing in the dial settings of rf equipment intended for single-sideband or independent-sideband transmission.

Assigned Frequency Band. The frequency band within which the emission of a station is authorized; the width of the band equals the necessary bandwidth plus twice the absolute value of the frequency tolerance. Where space stations are concerned, the assigned frequency band includes twice the maximum Doppler shift that may occur in relation to any point of the Earth=s surface.

Associated Common-channel Signaling. Common-channel signaling in which the signal channel is associated with a specific trunk group and terminates at the same pair of switches as the trunk group. *Note:* In associated common-channel signaling, the signaling is usually accomplished by using the same facilities as the associated trunk group.

Astable Multivibrator. An oscillator that produces a square wave output from a DC voltage.

Asynchronous. With reference to video and data signals and devices: not being precisely in step, not of the same frequency, or not happening together in time.

Asynchronous Transfer Mode (ATM). A high-speed multiplexing and switching method utilizing fixed-length cells of 53 octects to support multiple types of traffic. *Note:* ATM, specified in international standards, is asynchronous in the sense that cells carrying user data need not be periodic.

"AT" Instruction Set. ATtention: A standard modem control command-set protocol developed by modem-manufacturer Hayes, used by most modem manufacturers.

ATM. Asynchronous Transfer Mode. A network technology for both local and wide area networks (LANs and WANs) that supports real time voice and video as well as data. The topology uses switches that establish a logical circuit from end to end, which guarantees quality of service. However, unlike telephone switches that dedicate circuits end to end, unused bandwidth in ATM's logical circuits can be appropriated when needed. For example, idle bandwidth in a videoconference circuit can be used to transfer data.

Atmospheric Duct. A horizontal layer in the lower atmosphere in which the vertical refractive index gradients are such that radio signals (a) are guided or focused within the duct, (b) tend to follow the curvature of the Earth, and (c) experience less attenuation in the ducts than they would if the ducts were not present. *Note:* The reduced refractive index at the higher altitudes bends the signals back toward the Earth. Signals in a higher refractive index layer, *i.e.,* duct, tend to remain in that layer because of the reflection and refraction encountered at the boundary with a lower refractive index material.

Atom. The smallest particle that an element can be broken down into and still maintain its unique identity.

Atomic Number. The number of positive charges or protons in the nucleus of an atom.

AT&T. American Telephone & Telegraph Company, a major US common carrier.

Attention Signal. The attention signal to be used by AM, FM, and TV broadcast stations to actuate muted receivers for inter-station receipt of emergency cuing announcements and broadcasts involving a range of emergency contingencies posing a threat to the safety of life or property.

Attenuate. To reduce the amplitude of an action or signal. The opposite of amplification.

Attenuation. The decrease in intensity of a signal, beam, or wave as a result of absorption of energy and of scattering out of the path to the detector, but not including the reduction due to geometric spreading. *Note 1:* Attenuation is usually expressed in dB. *Note 2: "Attenuation"* is often used as a misnomer for "attenuation coefficient," which is expressed in dB per kilometer. *Note 3:* A distinction must be made as to whether the attenuation is that of signal power or signal electric field strength.

Attenuator. 1. In electrical systems, a network that reduces the amplitude of a signal without appreciably distorting its waveform. *Note 1:* Electrical attenuators are usually passive devices. *Note 2:* The degree of attenuation may be fixed, continuously adjustable, or incrementally adjustable. Fixed attenuators are often called pads, especially in telephony. *Note 3:* The input and output impedances of an attenuator are usually matched to the impedances of the signal source and load, respectively. 2. In optical systems, a device that reduces the amplitude of a signal without appreciably distorting its waveform. *Note 1:* Optical attenuators are usually passive devices. *Note 2:* The degree of attenuation may be fixed, continuously adjustable, or incrementally adjustable.

Audio. Relating to frequencies that can be heard by the human ear. Approximately 20 Hz. to 20 kHz.

Audio Response Unit (ARU). A device that provides synthesized voice responses to dual-tone multifrequency signaling input by processing calls based on (a) the call-originator input, (b) information received from a host data base, and (c) information in the incoming call, such as the time of day. *Note:* ARUs are used to increase the number of information calls handled and to provide consistent quality in information retrieval.

Automatic Frequency Control (AFC). A device or circuit that maintains the frequency of an oscillator within the specified limits with respect to a reference frequency.

Automatic Link Establishment (ALE). 1. In high-frequency (HF) radio, the capability of a station to make contact, or initiate a circuit, between itself and another specified radio station, without human intervention and usually under processor control. *Note:* ALE techniques include automatic signaling, selective calling, and automatic handshaking. Other automatic techniques that are related to ALE are channel scanning and selection, link quality analysis (LQA), polling, sounding, message store-and-forward, address protection, and anti-spoofing. 2. In HF radio, a link control system that includes automatic scanning, selective calling, sounding, and transmit channel selection using link quality analysis data. *Note:* Optional ALE functions include polling and the exchange of orderwire commands and messages.

Autonomous System. A stand-alone PV system that has no back-up generating source. May or may not include storage batteries. Most battery systems are designed for a certain minimum "days of autonomy" — which means that the batteries can supply sufficient power with no sunlight to charge the batteries. This varies from 3-5 days in the sunbelt, to 5 to 10 days elsewhere.

Autotransformer. A single winding transformer where the output is taken from taps on the winding.

Auxiliary Power. Electric power that is provided by an alternate source and that serves as backup for the primary power source at the station main bus or prescribed sub-bus. *Note 1:* An offline unit provides electrical isolation between the primary power source and the critical technical load whereas an online unit does not. *Note 2:* A Class A power source is a primary power source, *i.e.,* a source that assures an essentially continuous supply of power. *Note 3:* Types of auxiliary power services include Class B, a standby power plant to cover extended outages of the order of days; Class C, a 10-to-60-second quick-start unit to cover short-term outages of the order of hours; and Class D, an uninterruptible non-break unit using stored energy to provide continuous power within specified voltage and frequency tolerances.

Availability. 1. Describes the reliability of power plants. It refers to the number of hours the turbines are available to produce power divided by the total hours in a year.

2. A numeric representation of the period of time that a transmission facility performs better than some minimum performance value, usually expressed as a percentage of time on an annual basis. Performance values are usually specified based on bit-error-rate (BER) thresholds for digital channels and weighted signal-to-noise ratio for analog channels. High-quality networks will have an availability of about 99.96% or better. The statistical distribution of outages and degraded performance depend upon the nature of the transmission media involved.

Avalanche Multiplication. A current-multiplying phenomenon that occurs in a semiconductor photodiode that is reverse-biased just below its breakdown voltage. *Note:* Under such a condition, photocurrent carriers, *i.e.*, electrons, are swept across the junction with sufficient energy to ionize additional bonds, creating additional electron-hole pairs in a regenerative action.

Average Value. A value of voltage or current where the area of the wave above the value equals the area of the wave below the value.

AVC. Abbreviation for "automatic volume control"

Avionics. Aviation electronics.

Avoided Cost. The minimum amount an electric utility is required to pay an independent power producer, under the PURPA regulations of 1978, equal to the costs the utility calculates it avoids in not having to produce that power (usually substantially less than the retail price charged by the utility for power it sells to customers).

AVT. AudioVisual Terminal.

AWG. American Wire Gauge, a standard system for designating the size of electrical wire. The higher the number, the smaller the wire. Most house wiring is #12 or 14. In most other countries, wire is specified by the size in millimeters.

Azimuth. Angle between the north direction and the projection of the surface normal into the horizontal plane; measured clockwise from north. As applied to the PV array, 180 degree azimuth means the array faces due south.

B

B8ZS. Bipolar with 8 Zeros Substitution. A method for 1's density control for bipolar transmission signals such as T1. See also *D4*.

Backhaul. In wireless technology, backhaul refers to transporting voice and data traffic from a cell site to the switch.

Backscattering. 1. Radio wave propagation in which the direction of the incident and scattered waves, resolved along a reference direction (usually horizontal) are oppositely directed. A signal received by backscattering is often referred to as "backscatter." 2. In optics, the scattering of light into a direction generally opposite to the original one.

Back-to-back Connection. 1. A direct connection between the output of a transmitting device and the input of an associated receiving device. *Note:* When used for equipment measurements or testing purposes, such a back-to-back connection eliminates the effects of the transmission channel or medium. 2. A direct connection between the output of a receiving device and the input to a transmitting device. *Note:* The term *"direct,"* as used in both definitions, may be construed as permitting a passive device such as a pad (attenuator) to accommodate power level constraints.

Backward Channel. 1. In data transmission, a secondary channel in which the direction of transmission is constrained to be opposite to that of the primary, *i.e.*, the forward (user-information) channel. *Note:* The direction of transmission in the backward channel is restricted by the control interchange circuit that controls the direction of transmission in the primary channel. 2. In a data circuit, the channel that passes data in a direction opposite to that of its associated forward channel. *Note 1:* The backward channel is usually used for transmission of supervisory, acknowledgement, or error-control signals. The direction of flow of these signals is opposite to that in which user information is being transferred. *Note 2:* The backward-channel bandwidth is usually less than that of the primary channel, *i.e.*, the forward (user information) channel.

Bag Phone. An older type of phone, more technically referred to as a transportable, that provides the advantages of an A/C adaptor for unlimited power, an external antenna for better coverage and greater output that creates a stronger signal than a portable phone can offer.

Balanced Bridge. Condition that occurs when a bridge circuit is adjusted to produce a zero output.

Balanced Code. 1. In PCM systems, a code constructed so that the frequency spectrum resulting from the transmission of any code word has no dc component. 2. In PCM, a code that has a finite digital sum variation.

Balanced Line. A transmission line consisting of two conductors in the presence of ground, capable of being operated in such a way that when the voltages of the two conductors at all transverse planes are equal in magnitude and opposite in polarity with respect to ground, the currents in the two conductors are equal in magnitude and opposite in direction. *Note:* A balanced line may be operated in an unbalanced condition. *Synonym* balanced signal pair.

Balanced Modulator. A modulator constructed so that the carrier is suppressed and any associated carrier noise is balanced out. *Note 1:* The balanced modulator output contains only the sidebands. *Note 2:* Balanced modulators are used in AM transmission systems.

Balance of System. Represents all components and costs other than the PV modules. It includes design costs, land, site preparation, system installation, support structures, power conditioning, operation and maintenance costs, indirect storage, and related costs.

Balancing Network. 1. In a hybrid set, hybrid coil, or resistance hybrid, a circuit used to match, *i.e.*, to balance, the impedance of a uniform transmission line, *i.e.*, twisted metallic pair, over a selected range of frequencies. *Note:* A balancing network is required to ensure isolation between the two ports of the four-wire side of the hybrid. 2. A device used between a balanced device or line and an unbalanced device or line for the purpose of transforming from balanced to unbalanced or from unbalanced to balanced.

Band Gap. In a semiconductor, the energy difference between the highest valence band and the lowest conduction band.

Band Gap Energy (Eg). The amount of energy (in electron volts) required to free an outer shell electron from its orbit about the nucleus to a free state and, thus, to promote it from the valence level to the conduction level.

Band-pass Filter. 1. A tuned circuit designed to pass a band of frequencies between a lower cut-off frequency (f_1) and a higher cut-off frequency (f_2). Frequencies above and below the pass band are heavily attenuated.

2. A filter that ideally passes all frequencies between two non-zero finite limits and bars all frequencies not within the limits. *Note:* The cutoff frequencies are usually taken to be the 3-dB points.

Band-stop Filter. 1. A tuned circuit designed to stop frequencies between a lower cut-off frequency (f_1) and a higher cut-off frequency (f_2) of the amplifier while passing all other frequencies.

2. A filter that attenuates, usually to very low levels, all frequencies between two non-zero, finite limits and passes all frequencies not within the limits. *Note:* A band-stop filter

may be designed to stop the specified band of frequencies but usually only attenuates them below some specified level. *Synonyms* band-elimination filter, band-rejection filter, band-suppression filter, notched filter.

Band-to-band Auger Recombination. Recombination of an electron and a hole occurring between bands of the same energy in which no magnetic radiation is emitted.

Bandwidth. 1. Width of the band of frequencies between the half power points. 2. A relative range of frequencies that can carry a signal without distortion on a transmission medium. Sometimes referred to as a 'pipe.' 3. The difference between the highest and lowest frequencies passed by an analog communications channel, device or system.

Bandwidth (of an optical fiber). 1. The lowest modulation frequency at which the RMS peak-to-valley amplitude (optical power) difference of an intensity-modulated monochromatic signal decreases, at the output of the fiber, to a specified fraction (usually one-half) of the RMS peak-to-valley amplitude (optical power) difference of a nearly-zero (arbitrarily low) modulation frequency, both modulation frequencies having the same RMS peak-to-valley amplitude (optical power) difference at the fiber input. *Note 1:* In multimode fibers, multimode distortion is usually the most significant parameter limiting fiber bandwidth, although material dispersion may also play a significant role, especially in the first (850-nm) window. *Note 2:* In multimode fibers, the bandwidth$distance product (colloquially, *"fiber bandwidth"*) is customarily specified by vendors for the bandwidth as limited by multimode distortion only. The spectral width of the optical source is assumed to be extremely narrow. In practice, the effective fiber bandwidth will also be limited by dispersion, especially in the first (850-nm) window, where material dispersion is relatively high, because optical sources have a finite spectral width. Laser diodes typically have a spectral width of several nanometers, FWHM. LEDs typically have a spectral width of 35 to 100 nm, FWHM. *Note 3:* The effective risetime of multimode fibers may be estimated fairly accurately as the square root of the sum of the squares of the material-dispersion-limited risetime and the multimode-distortion-limited risetime. *Note 4:* In single-mode fibers, the most important parameters affecting fiber bandwidth are material dispersion and waveguide dispersion. Practical fibers are designed so that material dispersion and waveguide dispersion cancel one another at the wavelength of interest. *Note 5:* Regarding effective fiber bandwidth as it affects overall system performance, it should be recognized that optical detectors such as PIN diodes are square-law devices. Their photocurrent is proportional to the optical power of the detected signal. Because electrical power is a function of the square of the current, when the optical power decreases by one-half (a 3-dB decrease), the electrical power decreases by three-fourths (a 6-dB decrease). 2. Loosely, *synonym* bandwidth$distance product.

Barrier Energy. The energy given up by an electron in penetrating the cell barrier; a measure of the electrostatic potential of the barrier.

Barrier Potential. The natural difference of potential that exists across a forward biased pn junction.

BAS. Bit-rate Allocation Signal. In Rec. H.221, T.120 *et al*, a data word within the frame structure used to transmit commands, control and indication signals and capabilities.

Base. The region that lies between the emitter and collector of a bipolar junction transistor (BJT).

Baseband. 1. The original band of frequencies produced by a transducer, such as a microphone, telegraph key, or other signal-initiating device, prior to initial modulation. *Note 1:* In transmission systems, the baseband signal is usually used to modulate a carrier. *Note 2:* Demodulation re-creates the baseband signal. *Note 3: Baseband* describes the signal state prior to modulation, prior to multiplexing, following demultiplexing, and following demodulation. *Note 4:* Baseband frequencies are usually characterized by being much lower in frequency than the frequencies that result when the baseband signal is used to modulate a carrier or subcarrier. 2. In facsimile, the frequency of a signal equal in bandwidth to that between zero frequency and maximum keying frequency.

Base Biasing. A method of biasing a BJT in which the bias voltage is supplied to the base by means of a resistor.

Baseline Performance Value. Initial values of Isc, Voc, Pmp, Imp measured by the accredited laboratory and corrected to Standard Test Conditions, used to validate the manufacturer's performance measurements provided with the qualification modules per IEEE 1262.

Base Power. Power generated by a utility unit that operates at a very high capacity factor.

Base Station Controller. The part of the wireless system's infrastructure that controls one or multiple cell sites' radio signals, thus reducing the load on the switch. It can be viewed as a form of distributed processing.

Base Transceiver Station. (BTS): The portion of the wireless system's infrastructure that is responsible for sending and receiving the actual radio signals over the airwaves. This device takes radio signals from subscribers' phones and sends them over leased telephone lines or microwave signals to the switch.

Battery. A DC voltage source containing two or more cells that convert chemical energy to electrical energy.

Battery Energy Storage. The three main applications for battery energy storage systems include spinning reserve at generating stations, load leveling at substations, and peak shaving on the customer side of the meter. Battery storage has also been suggested for holding down air emissions at the power plant by shifting the time of day of the emission or shifting the location of emissions.

Baud (Bd). 1. A unit of modulation rate. *Note:* One baud corresponds to a rate of one unit interval per second, where the modulation rate is expressed as the reciprocal of the duration in seconds of the shortest unit interval. 2. A unit of signaling speed equal to the number of discrete signal conditions, variations, or events per second. *Note 1:* If the duration of the unit interval is 20 milliseconds, the signaling speed is 50 bauds. If the signal transmitted during each unit interval can take on any one of *n* discrete states, the bit rate is equal to the rate in bauds times $\log_2 n$. The technique used to encode the allowable signal states may be any combination of amplitude, frequency, or phase modulation, but it cannot use a further time-division multiplexing technique to subdivide the unit intervals into multiple subintervals. In some signaling systems, non-information-carrying signals may be inserted to facilitate synchronization; *e.g.*, in certain forms of binary modulation coding, there is a forced inversion of the signal state at the center of the bit interval. In these cases, the synchronization signals are included in the calculation of the rate in bauds but not in the computation of bit rate. *Note 2: Baud* is sometimes used as a synonym for *bit-per-second*. This usage is deprecated.

B-Band Cellular. In the U.S. cellular duopoly, the regional Bell operating company's cellular subsidiary.

BBC. British Broadcasting Corporation.

B-CDMA (Broadband Code Division Multiple Access). A technology developed by InterDigital Communications Corp.

BCH (Bose-Chaudhuri-Hocquenghem). A low-overhead forward error correction technique often used for videoconferencing as in H.261 *et al*; especially effective for correcting single (non-burst) errors.

Beam Diameter. Of an electromagnetic beam, along any specified line that (a) intersects the beam axis and (b) lies in any specified plane normal to the beam axis, the distance between the two diametrically opposite points at which the irradiance is a specified fraction, *e.g.*, 1/2 or 1/ε, of the beam=s peak irradiance. *Note 1:* Beam diameter is usually used to characterize electromagnetic beams in the optical regime, and occasionally in the microwave regime, *i.e.*, cases in which the aperture from which the beam emerges is very large with respect to the wavelength. *Note 2:* Beam diameter usually refers to a beam of circular cross section, but not necessarily so. A beam may, for example, have an elliptical cross section, in which case the orientation of the beam diameter must be specified, *e.g.*, with respect to the major or minor axis of the elliptical cross section.

Beam Steering. Changing the direction of the main lobe of a radiation pattern. *Note:* In radio systems, beam steering may be accomplished by switching antenna elements or by changing the relative phases of the rf signals driving the elements. In optical systems, beam steering may be accomplished by changing the refractive index of the medium through which the beam is transmitted or by the use of mirrors or lenses.

Bel (B). A unit of measure of ratios of power levels, *i.e.*, relative power levels. *Note 1:* The number of bels for a given ratio of power levels is calculated by taking the logarithm, to the base 10, of the ratio. Mathematically, the number of bels is calculated as $B = \log_{10}(P_1/P_2)$ where P_1 and P_2 are power levels. *Note 2*: The dB, equal to 0.1 B, is a more commonly used unit.

Bell System. A group of affiliated RBOCs (Regional Bell Operating Companies) in the USA, which operate under consistent rules and specifications, many of which are set by AT&T.

Bent Pipe Technology. Satellite technology to transmit calls from one point on Earth to a satellite and back down to another point.

BER. Bit Error Rate, the absolute number of errors that may be present, often expressed as $FFx10^{-y}$, where FF is a floating point value and y is an exponent. For example, $3.456x10^{-8}$ See also *Bit Error Rate*.

Beta. (b) The ratio of collector current to base current in a bipolar junction transistor (BJT).

Bias. 1. A systematic deviation of a value from a reference value. 2. The amount by which the average of a set of values departs from a reference value. 3. Electrical, mechanical, magnetic, or other force (field) applied to a device to establish a reference level to operate the device. 4. In telegraph signaling systems, the development of a positive or negative dc voltage at a point on a line that should remain at a specified reference level, such as zero. *Note:* A bias may be applied or produced by (i) the electrical characteristics of the line, (ii) the terminal equipment, and (iii) the signaling scheme. 5. A DC voltage applied to a device to control its operation.

Bidding Credits. Discount sometimes given to small businesses in FCC spectrum auctions.

Big LEO. Low-earth orbit satellite system that will offer voice and data services; e.g., Iridium, Globalstar.

Billboard Antenna. An array of parallel dipole antennas with flat reflectors, usually positioned in a line or plane. *Note 1:* The spacing and dimensions of the dipoles depend on the wavelength. *Note 2:* The main lobe of a fixed billboard antenna may, within limits, be steered by appropriate phasing of the respective signals to individual elements of the array. *Synonym* broadside antenna.

Binary. A number system having only two symbols, 0 and 1. A base 2 number system.

Bipolar Junction Transistor (BJT). A three terminal device in which emitter to collector current is controlled by base current.

BIPV (Building-Integrated PV). A term for the design and integration of PV into the building envelope, typically replacing conventional building materials. This integration may be in vertical facades, replacing view glass, spandrel glass, or other facade material; into semitransparent skylight systems; into roofing systems, replacing traditional roofing materials; into shading "eyebrows" over windows; or other building envelope systems.

Bistable Multivibrator. A multivibrator with two stable states. An external signal is required to change the output from one state to the other. Also called a latch.

Bit. BInary digiT, the smallest unit of data. In uncompressed or uncoded data, it represents a binary 1 or 0.

Bit Error Rate (BER). A measure of quality of digital transmission systems. Expresses the ratio of correct bits to erroneous bits. To overcome transmission errors, many codecs include *forward error correction*, important for accurate recovery of transmitted, compressed data.

Bleeder Current. A current drawn continuously from a source. Bleeder current is used to stabilize the output voltage of a source.

Blocked Calls. An uncompleted call made from a wireless phone. Calls can be blocked for numerous reasons, but this typically refers to an instance where there are insufficient channels in a cell to handle the load of calls required. When a call is attempted within that cell and no channels are available, the call is "blocked" and a fast busy signal is heard by the subscriber.

Blocking Diode. A diode used to restrict or block reverse current from flowing backward through a module. [UL 1703] Alternatively, diode connected in series to a PV string; it protects its modules from a reverse power flow and, thus, against the risk of thermal destruction of solar cells.

Bluetooth. The code name for a new wireless technology being developed by Ericsson Inc., Intel Corp., Nokia Corp. and Toshiba. The technology enables data connections between electronic devices such as desktop computers, wireless phones, electronic organizers and printers in the 2.4 GHz range. Bluetooth would replace cable or infrared connections for such devices.

BNC. Bayonet Neill-Concelman. A type of bayonet-lock coaxial cable connector used for video and telecommunications.

BOC. *Bell Operating Company.*

Bode Plot. A graph of gain versus frequency.

Boltzmann's Constant (k). The number that relates the average energy of a molecule to its absolute temperature. *Note:* Boltzmann=s constant is approximately 1.38 x 10[B23] J/K (joules/kelvin).

Bonding. 1. In electrical engineering, the process of connecting together metal parts so that they make low resistance electrical contact for direct current and lower frequency alternating currents. 2. The process of establishing the required degree of electrical continuity between two or more conductive surfaces that are to be joined.

Boron (B). A chemical element, atomic number 5, semi-metallic in nature, used as a dopant to make p-semiconductor layers.

Boule. A sausage-shaped synthetic single-crystal mass grown in a special furnace, pulled and turned at a rate necessary to maintain the single-crystal structure during growth.

Bound Mode. In an optical fiber, a mode that (a) has a field intensity that decays monotonically in the transverse direction everywhere external to the core and (b) does not lose power to radiation. *Note:* Except for single-mode fibers, the power in bound modes is predominantly contained in the core of the fiber. *Synonyms* guided mode, trapped mode.

bps. Bits Per Second, a measurement of data rate, generally excluding the contributions or effects of error correction, encryption, framing, synchronization and similar such signals.

BRA. European term for *BRI.*

Branch Current. The portion of total current flowing in one path of a parallel circuit.

Braid. 1. An essential part of many fiber-optic cable designs, consisting of a layer of woven yarn. *Note:* In the case of single-fiber loose-buffered or two-fiber "zip-cord" loose-buffered fiber-optic cables, the braid is situated between the buffer tube and jacket. In the case of cables having multiple buffer tubes, the braid is usually situated between the inner jacket and outer jacket. 2. Loosely, an unwoven parallel bundle of yarn situated around the tight buffer of a single-fiber or two-fiber "zip-cord" fiber-optic cable. *Note 1:* The braid serves to add tensile strength to the cable. The braid may also be anchored to an optical connector or splice organizer assembly to secure the end of the cable. *Note 2:* The braid is often of an aramid yarn.

Breadboard. 1. An assembly of circuits or parts used to prove the feasibility of a device, circuit, system, or principle with little or no regard to the final configuration or packaging of the parts. 2. To prepare a breadboard.

Breakdown Voltage. Voltage at which the breakdown of a dielectric or insulator occurs.

Break-out Box. A testing device that permits a user to access individual leads of an interface cable, using jumper wires, in order to monitor, switch, or patch the electrical output of the cable.

Breakout Cable. A multifiber fiber-optic cable design in which individual fibers, usually tight-buffered, are surrounded by separate strength members and jackets, which are in turn enveloped by a common jacket. *Note 1:* The breakout cable facilitates easy installation of fiber-optic connectors. All that need be done to prepare the ends of the cable to receive connectors is to remove the outer jacket, exposing what are essentially individual single-fiber cables. *Note 2:* Because it tends to induce bends in the fibers, the breakout cable design usually results in slightly higher transmission losses, for a given fiber, than loose-buffer designs. *Synonym* fanout cable.

Breakover Voltage. Minimum voltage required to cause a diac to break down and conduct.

Breezeway. That portion of a composite video signal between the end of the trailing edge of each horizontal synchronization pulse and the beginning of the color reference burst signal (television communications usage).

BREW. Binary Runtime Environment for Wireless. An open-platform for wireless applications development, device configuration, application distribution, billing and payment, developed and licensed by Qualcomm.

BRI. *ISDN* Basic Rate Interface, 2B+D. In the US, B may equal 56- or 64-Kbps. The D or signaling channel is 16 Kbps.

Bridge Rectifier. A circuit using four diodes to provide full wave rectification. Converts an AC voltage to a pulsating DC voltage.

Brillouin Scattering. In a physical medium, scattering of lightwaves, caused by thermally driven density fluctuations. *Note:* Brillouin scattering may cause frequency shifts of several gigahertz at room temperature.

British Thermal Unit (Btu). The amount of heat energy required to raise the temperature of one pound of water from 60 degrees F to 61 degrees F at one atmosphere pressure.

Broadband. Using a wide-bandwidth channel for voice, data and/or video services.

Broadband PCS. Synonymous with personal communications services created in the A- through F-Block auctions and used for voice and data.

BT. *Bluetooth;* also the product of: symbol rate, Gaussian filter and bandwidth; used in DECT and other Bluetooth devices.

BTA (Basic Trading Area). Usually composed of several contiguous counties. BTAs are a service area designed by Rand McNally and adopted by the FCC. There are 493 BTAs in the United States.

Buffer. An amplifier used to isolate a load from a source.

Build-to-suit. The process by which independent firms find and build antenna sites to meet a carrier's specifications.

Bulk Resistance. The natural resistance of a "P" type or "N" type semiconductor material.

Bundling. Grouping various telecommunications services—wireline and/or wireless—as a package to increase the appeal to potential customers and reduce advertising, marketing and other expenses associated with delivering multiple services. For example, a bundled package could include long distance, cellular, Internet and paging services.

Burst Switching. In a packet-switched network, a switching capability in which each network switch extracts routing instructions from an incoming packet header to establish and maintain the appropriate switch connection for the duration of the packet, following which the connection is automatically released. *Note:* In concept, burst switching is similar to connectionless mode transmission, but it differs from the latter in that burst switching implies an intent to establish the switch connection in near real time so that only minimum buffering is required at the node switch.

Butterworth Filter. A type of active filter characterized by a constant gain (flat response) across the midband of the circuit and a 20 dB per decade roll-off rate for each pole contained in the circuit.

BW. Abbreviation for bandwidth.

B-Y. Blue minus luminance (Y). One of two color difference signals used in encoding NTSC composite television signals.

Bypass Capacitor. A capacitor used to provide an AC ground at some point in a circuit.

Bypass Diode. A diode connected across one or more solar cells in a photovoltaic module such that the diode will conduct if the cell(s) become reverse biased. [UL 1703] Alternatively, diode connected anti-parallel across a part of the solar cells of a PV module. It protects these solar cells from thermal destruction in case of total or partial shading of individual solar cells while other cells are exposed to full light.

Byte. Originally coined as a data chunk of from 1- to 6-bits, later up to 9-bits, spelled with a "y" to avoid confusion with bit. In modern usage, a byte is 8-bits, like *octet*, the ITU term which has always represented 8-bits.

C

Cable. 1. Group of two or more insulated wires. 2. An assembly of one or more insulated conductors, or optical fibers, or a combination of both, within an enveloping jacket. *Note:* A cable is constructed so that the conductors or fibers may be used singly or in groups. 3. A message sent by cable, or by any means of telegraphy.

*****Cable Cutoff Wavelength (ë$_{cc}$).** For a cabled single-mode optical fiber under specified length, bend, and deployment conditions, the wavelength at which the fiber=s second order mode is attenuated a measurable amount when compared to a multimode reference fiber or to a tightly bent single-mode fiber.

CAD. 1. Abbreviation for "computer aided design". 2. (Computer-aided Dispatch) Computer systems to help dispatch personnel and vehicles, commonly used by public-safety agencies.

Cadmium (Cd). A chemical element, atomic number 48, used in making certain types of solar cells and batteries.

Cadmium Telluride (CdTe). A polycrystalline thin-film photovoltaic material.

CALEA (Communications Assistance to Law Enforcement Act). A 1994 law granting law enforcement agencies the ability to wiretap new digital networks and requiring wireless and wireline carriers to enable eavesdropping equipment use in digital networks.

Calibration. To adjust the correct value of a reading by comparison to a standard.

Call Diversion. The term used for call forwarding on a GSM system.

Call Forwarding. A feature on both wired and wireless phones that sends calls to another number from that which was originally called. A typical example would be to forward the calls from your office phone to your wireless phone when you are away from your office.

Call Management. 1. In telegraphy, route selection, signaling, and circuit usage and availability for a call. 2. In universal personal telecommunications, the ability of a user to inform the network how to handle incoming calls in accord with certain parameters, such as the call originator, the time of day, and the nature of the call. *Note:* Call management is accomplished by means of information in the user=s service profile.

Call Set-up Time. 1. The overall length of time required to establish a circuit-switched call between users. 2. For data communication, the overall length of time required to establish a circuit-switched call between terminals; *i.e.*, the time from the initiation of a call request to the beginning of the call message. *Note:* Call set-up time is the summation of: (a) call request timeCthe time from initiation of a calling signal to the delivery to the caller of a proceed-to-select signal; (b) selection timeCthe time from the delivery of the proceed-to-select signal until all the selection signals have been transmitted; and (c) post selection timeCthe time from the end of the transmission of the selection signals until the delivery of the call-connected signal to the originating terminal.

Call Timers. A feature on wireless phones that shows the user how long the current call has progressed, or how many minutes have been used cumulatively since the last time the "Total Call Timer" feature has been reset.

Caller ID. An enhanced feature that displays a caller's phone number on the wireless handset.

Calling Party Pays. This service bills the originator of a call to a wireless device rather than the receiver and is more common in other countries than in the United States. However, many U.S. carriers are pushing for calling party pays, since it would probably increase minutes of use.

Capacitance. The ability of a capacitor to store an electrical charge. The basic unit of capacitance is the Farad.

Capacitive Coupling. The transfer of energy from one circuit to another by means of the mutual capacitance between the circuits. *Note 1:* The coupling may be deliberate or inadvertent. *Note 2:* Capacitive coupling favors transfer of the higher frequency components of a signal, whereas inductive coupling favors lower frequency components, and conductive coupling favors neither higher nor lower frequency components.

Capacitive Reactance. The opposition to current flow provided by a capacitor. Capacitive reactance is measured in ohms and varies inversely with frequency.

Capacitor. An electronic component having capacitive reactance.

Capacitor Microphone. Microphone whose operation depends on variations in capacitance caused by varying air pressure on the movable plate of a capacitor.

Capacity Factor. The amount of energy that the system produces at a particular site as a percentage of the total amount that it would produce if it operated at rated capacity during the entire year. For example, the capacity factor for a wind farm ranges from 20% to 35%. Thirty-five percent is close to the technology potential.

Cap Code. A pager's unique electronic identification number on the back of the device

Capture Effect. A phenomenon, associated with FM reception, in which only the stronger of two signals at or near the same frequency will be demodulated. *Note 1:* The complete suppression of the weaker signal occurs at the receiver limiter, where it is treated as noise and rejected. *Note 2:* When both signals are nearly equal in strength, or are fading independently, the receiver may switch from one to the other. *Synonym* FM capture effect.

Carbon-film Resistor. Device made by depositing a thin carbon film on a ceramic form.

Carbon Microphone. Microphone whose operation depends on pressure variation in carbon granules causing a change in resistance.

Carbon Resistor. Resistor of fixed value made by mixing carbon granules with a binder which is moulded and then baked.

Cardinal Radial. Those eight radials at 0E, 45E, 90E, 135E, 180E, 225E, 270E, and 315E of azimuth with respect to true north. *Note:* The four radials at 0E, 90E, 180E, and 270E of azimuth with respect to true north are referred to as the cardinal points. The cardinal points are equivalent to true north, east, south, and west.

Carrier Power (of a radio transmitter). 1. The average power supplied to the antenna transmission line by a transmitter during one radio frequency cycle taken under the condition of no modulation. *Note:* The concept does not apply to pulse modulation or frequency-shift keying. 2. The average unmodulated power supplied to a transmission line.

Carrier Shift. 1. In the transmission of binary or teletypewriter signals, keying in which the carrier frequency is shifted in one direction for marking signals and in the opposite direction for spacing signals. 2. In amplitude modulation, a condition that results from imperfect modulation in which the positive and negative excursions of the modulating envelope are unequal in amplitude. *Note 1:* The carrier shift results in a change in carrier power. *Note 2:* The carrier shift may be a shift to a higher or to a lower frequency.

Carrier-to-receiver Noise Density (*C*/k*T*). In satellite communications, the ratio of the received carrier power to the receiver noise power density. *Note 1:* The carrier-to-receiver noise density ratio is usually expressed in dB. *Note 2:* The carrier-to-receiver noise density is given by *C*/k*T*, where C is the received carrier power in watts, k is Boltzmann=s constant in joules per kelvin, and *T* is the receiver system noise temperature in kelvins. *Note 3:* The receiver noise power density, k*T*, is the receiver noise power per hertz.

Cascaded Amplifier. An amplifier with two or more stages arranged in a series configuration.

Cascode Amplifier. A high frequency amplifier made up of a common-source amplifier with a common-gate amplifier in its drain network.

Category 3. The ANSI/EIA/TIA-568 designation for 100-ohm unshielded twisted-pair cables and associated connecting hardware whose characteristics are specified for data transmission up to 16 Mb/s.

Category 4. The ANSI/EIA/TIA-568 designation for 100-ohm unshielded twisted-pair cables and associated connecting hardware whose characteristics are specified for data transmission up to 20 Mb/s.

Category 5. The ANSI/EIA/TIA-568 designation for 100-ohm unshielded twisted-pair cables and associated connecting hardware whose characteristics are specified for data transmission up to 100 Mb/s.

Cathode. The negative terminal electrode of a device. The "N" material in a junction diode.

Cathode Ray Tube (CRT). Vacuum tube used to display data in a visual form. Picture tube of a television or computer terminal.

Cathodic Protection. A method of preventing oxidation (rusting) of exposed metal structures, such as bridges and pipelines, by imposing between the structure and the ground a small electrical voltage that opposes the flow of electrons and that is greater than the voltage present during oxidation.

CATV. *Community Antenna TeleVision* (system), commonly known as "cable-TV".

Cavity. A volume defined by conductor-dielectric or dielectric-dielectric reflective boundaries, or a combination of both, and having dimensions designed to produce specific interference effects (constructive or destructive) when excited by an electromagnetic wave.

C-band. 1. A microwave frequency band. When used for satellite links, the frequencies assigned are: uplink (to the satellite): 5.925-6.425 GHz. downlink (from the satellite): 3.700-4.200 GHz. C-band signals are only slightly attenuated by rainfall, making its use advantageous over Ku-band for some applications.

2. *Colloquially,* a frequency band between 4 GHz and 6 GHz used in satellite communications. *Note 1:* For procurement purposes, the radio frequency band(s) must be specified using the upper and lower limits of the band, per 47 CFR 300. *Note 2:* Letter designators of radio frequency bands are imprecise, deprecated, and legally obsolete.

CCIR (now ITU-R). From the French for the International Radio Consultative Committee, an International Telecommunications Union (ITU) (United Nations) body that mainly sets international standards for radio and satellite telecommunications.

CCIR 601. A CCIR recommendation regarding various standardized image resolutions for coding high quality video images, primarily for broadcasting. It provides the foundation for MPEG1 and MPEG2 resolutions and coding.

CCIS. Common Channel Inter-office Signaling. A technique where signaling for a number of channels is carried on a single common channel, in contrast to each channel carrying its own signaling information.

CCITT (now ITU-T). From the French for the International Telegraph and Telephone Consultative Committee. An International Telecommunications Union (United Nations) body that sets international standards for telecommunications.

CCR. *Comparison Category Rating.* An ITU-T P.800 subjective test where the signal from a device under test is compared to a known-quality signal and may be rated from better to worse. See also DCR. This is a test performed by HEI.

CCTV. *Closed Circuit TeleVision.* Non-broadcast television, intended for limited user access, often used for security and monitoring.

CDMA (Code Division Multiple Access). 1. A spread spectrum air interface technology used in some digital cellular, personal communications services and other wireless networks. 2. *Code Division Multiple Access*, a spread-spectrum technique used with Ku-band satellites to provide mobile data, to truckers for example. CDMA provides benefits under multipath conditions. Also, the IS-95 digital PCS cellular standard in the USA.

CDMA2000. Code division multiple access. CDMA2000 refers to current-generation, digital wireless network technology.

CDMA2000 1X. Code division multiple access. CDMA2000 1X is one type of IP-based, so-called 'third-generation' wireless technology, enabling wireless data transfer speeds of up to 144 kilobits per second.

cdmaOne (Code Division Multiple Access). The IS-95 CDMA standard developed by Qualcomm Inc.; a word coined by the CDMA Development Group.

CDPD (Cellular Digital Packet Data). An enhanced system overlay for transmitting and receiving data over cellular networks.

Cell. 1. Single unit used to convert chemical energy into a DC electrical voltage. 2. A geographic area within a wireless system that is covered by the signal sent and received by the transmitter and receiver equipment located within that area. Typically referred to as a "cell site," these are represented by hexagonal shapes by engineers when planning systems. That shape was originally derived from the honeycomb of bees, within which each single unit is referred to as a cell.

Cell Barrier. A very thin region of static electric charge along the interface of the positive and negative layers in a photovoltaic cell. The barrier inhibits the movement of electrons from one layer to the other, so that higher-energy electrons from one side diffuse preferentially through it in one direction, creating a current and thus a voltage across the cell. Also called depletion zone, cell junction, or space charge.

Cell Junction. The area of immediate contact between two layers (positive and negative) of a photovoltaic cell. The junction lies at the center of the cell barrier or depletion zone.

Cell Site. The location where the wireless antenna and network communications equipment is placed.

Cell Splitting. The process of creating more coverage and capacity in a geographic area by having more than one cell cover the same area that a single cell originally did. Each cell then covers a smaller area, with lower power, and thus offers the ability to reuse frequencies more times in a larger geographic coverage area such as a city or MTA.

Cellemetry. Brand name for Cellemetry LLC's telemetry service, which uses the cellular network to carry data messaging used for remote services such as utility meter reading, vending machine status and vehicle or trailer tracking.

Cellular. The name given to the original concept of dividing a large geographic area into smaller coverage areas called cells. Each cell handles calls on different channels and communicates with the central processing unit, called a switch, to facilitate the handing-off of calls from one cell to another as a user moves through the system. Cellular is currently used in hundreds of countries worldwide and boasts more than 200 million subscribers.

Center Clipper. In an Acoustic Echo Canceller (AEC) or a Telephony Echo Canceller (TEC), a clipper that only operates at very low signal levels, to remove residual echo and artifacts.

Center Frequency. Frequency to which an amplifier is tuned. The frequency half way between the cut-off frequencies of a tuned circuit.

Center Tap. Midway connection between the two ends of a winding.

Center Tapped Rectifier. Circuit that make use of a center tapped transformer and two diodes to provide full wave rectification.

Center Tapped Transformer. A transformer with a connection at the electrical center of a winding.

Central Power. The generation of electricity in large power plants with distribution through a network of transmission lines (grid) for sale to a number of users. Opposite of distributed power.

CEPT. Conference of European Post and Telecom administrations.

Ceramic Capacitor. Capacitor in which the dialectric is ceramic.

Cesium Standard. A primary frequency standard in which electronic transitions between the two hyperfine ground states of cesium-133 atoms is used to control the output frequency. *Note*: The energy level between the two hyperfine ground states corresponds, in the absence of external influences (*e.g.*, the magnetic field of the Earth), to a frequency of 9,192,631,770 Hz.

CFS. Communications Frame Structure.

CGI. *Common Gateway Interface*, a method of extending web server functionality using scripts or programs in response to Web browser requests.

Checksum. The information produced by a mathematical technique used to determine whether or not errors have been introduced into a data stream. Usually used for low-bit-rate data transmission. A checksum does not include *forward error correction* capabilities. Compare with *syndrome*.

Channel. 1. A connection between initiating and terminating nodes of a circuit. 2. A single path provided by a transmission medium via either (a) physical separation, such as by multipair cable or (b) electrical separation, such as by frequency- or time-division multiplexing. 3. A path for conveying electrical or electromagnetic signals, usually distinguished from other parallel paths. 4. Used in conjunction with a predetermined letter, number, or codeword to reference a specific radio frequency. 5. The portion of a storage medium, such as a track or a band, that is accessible to a given reading or writing station or head. 6. In a communications system, the part that connects a data source to a data sink.

Channel Noise Level. 1. The ratio of the channel noise at any point in a transmission system to an arbitrary level chosen as a reference. *Note 1:* The channel noise level may be expressed in (a) dB above reference noise (dBrn), (b) dB above reference noise with C-message weighting (dBrnC), or (c) adjusted dB (dBa). *Note 2:* Each unit used to measure channel noise level reflects a circuit noise reading of a specialized instrument designed to account for different interference effects that occur under specified conditions. 2. The noise power density spectrum in the frequency range of interest. 3. The average noise power in the frequency range of interest.

Chapter 11. The section of the federal Bankruptcy Code enabling distressed, but not necessarily insolvent, companies to seek protection from creditors while they reorganize their finances. The proceedings are governed by a bankruptcy court judge who must approve any plan of reorganization. Creditors can be represented in the proceedings and present alternative reorganization plans.

Chapter 7. The section of the federal Bankrutpcy Code governing liquidation of insolvent companies in order to pay their creditors.

Characteristic Distortion. In telegraphy, the distortion caused by transients that, as a result of previous modulation, are present in the transmission channel. *Note:* Characteristic distortion effects are not consistent. Their effects on a given signal transition are dependent upon transients remaining from previous signal transitions.

Characteristic Frequency. A frequency which can be easily identified and measured in a given emission. A carrier frequency may, for example, be designated as the characteristic frequency.

Characteristic Impedance (Z_o). 1. The impedance of a circuit that, when connected to the output terminals of a uniform transmission line of arbitrary length, causes the line to appear infinitely long. *Note 1:* A uniform line terminated in its characteristic impedance will have no standing waves, no reflections from the end, and a constant ratio of voltage to current at a given frequency at every point on the line. *Note 2:* If the line is not uniform, the iterative impedance must be used. 2. For Maxwell=s equations, the impedance of a linear, homogeneous, isotropic, dielectric propagation medium free of electric charge, given by the relation $Z=(\mu/\varepsilon)^{1/2}$ where μ is the magnetic permeability and ε is the electric permittivity of the medium.

Charge Back. The funds a carrier will 'charge back' to an agent or dealer if a customer discontinues service shortly after buying the product.

Charge. Quantity of electrical energy.

Charge Carrier. A free and mobile conduction electron or hole in a semiconductor.

Charge Controller. An electronic device which regulates the voltage applied to the battery system from the PV array. Essential for ensuring that batteries obtain maximum state of charge and longest life.

Charge Current. Current that flows to charge a capacitor or battery when voltage is applied.

Chassis. Metal box or frame into which components are mounted.

Chassis Ground. Connection to a chassis.

Chebyshev Filter. A type of active filter characterized by high roll-off rates (40 dB per decade per pole) and midband gain that is not constant.

Chemical Vapor Deposition (CVD). A method of depositing thin semiconductor films. With this method, a substrate is exposed to one or more vaporized compounds, one or more of which contain desirable constituents. A chemical reaction is initiated, at or near the substrate surface, to produce the desired material that will condense on the substrate.

Chip. 1. *Synonym* integrated circuit. 2. In satellite communications systems, the smallest element of data in an encoded signal. 3. The most elemental component of a spread spectrum signal when it is decompressed in time; that is, the longest duration signal in which signal parameters are approximately constant.. 4. In micrographic and display systems, a relatively small and separate piece of microform that contains microimages and coded information for search, identification, and retrieval purposes.

Chip Set. A combination of two or more integrated circuits on one module.

Chirping. 1. The rapid changing, as opposed to long-term drifting, of the frequency of an electromagnetic wave. *Note:* Chirping is most often observed in pulsed operation of a source. 2. A pulse compression technique that uses (usually linear) frequency modulation during the pulse.

Chlorofluorocarbon. A family of chemicals composed primarily of carbon, hydrogen, chlorine, and fluorine whose principal applications are that of refrigerants and industrial cleansers and whose principal drawback is the tendency to destroy the Earth's protective ozone layer.

Choke. Inductor used to oppose the flow of alternating current.

Chroma Keying. In television, nearly instantaneous switching between multiple video signals, based on the state, *i.e.*, phase, of the color (chroma) signal of one, to form a single composite video signal. *Note 1:* Chroma keying is used to create an overlay effect in the final picture, *e.g.*, to insert a false background, such as a weather map or scenic view, behind the principal subject being photographed. *Note 2:* The principal subject is photographed against a background having a single color or a relatively narrow range of colors, usually in the blue. When the phase of the chroma signal corresponds to the preprogrammed state or states associated with the background color, or range of colors, behind the principal subject, the signal from the alternate, *i.e.*, false, background is inserted in the composite signal and presented at the output. When the phase of the chroma signal deviates from that associated with the background color(s) behind the principal subject, video associated with the principal subject is presented at the output. *Synonym* color keying. *Colloquial synonym* blue-screening. *Contrast with* chrominance signal, composite video.

Chrominance. That portion of a composite video signal containing the color information (amplitude and phase information representing color-hue and saturation) in the image.

Churn. A measure of the number of subscribers who leave or switch to another carrier's service.

CIF. (Full) Common Intermediate Format. When used with ITU-T Rec. H.261, H.263 and other coding methods, the CIF describes the type of coded video frame transmitted.

The CIF is 352H x 288V pixels at video frame rates of 7.5, 10, 15 and 30 fps. Note that a codec providing no more than 7.5 fps is compliant with the recommendation. If better motion handling is needed, look for a numerically greater video frame rate specification.

The following table provides the range of resolutions possible under H.320 and H.323:

Picture Format	*Luminance Pixels*	*Mode*	
		H.261	*H.263*
SQCIF	128H x 96V	O	R
QCIF	176H x 144V	R	R
CIF	352H x 288V	O	O
4CIF	704H x 576V	-	O
16CIF	1408H x 1152V	-	O

where: S = Sub, Q = Quarter, H = Horizontal pixels, V = Vertical pixels, R = Required, O = Optional and, - = Not Defined.

Circuit. 1. The complete path between two terminals over which one-way or two-way communications may be provided. 2. An electronic path between two or more points, capable of providing a number of channels. 3. A number of conductors connected together for the purpose of carrying an electrical current. 4. An electronic closed-loop path among two or more points used for signal transfer. 5. A number of electrical components, such as resistors, inductances, capacitors, transistors, and power sources connected together in one or more closed loops.

Circuit Breaker. A protective device used to open a circuit when current exceeds a maximum value. In effect a reusable fuse.

Circuit Noise Level. At any point in a transmission system, the ratio of the circuit noise at that point to an arbitrary level chosen as a reference. *Note:* The circuit noise level is usually expressed in dBrn0, signifying the reading of a circuit noise meter, or in dBa0, signifying circuit noise meter reading adjusted to represent an interfering effect under specified conditions.

Circular Polarization. In electromagnetic wave propagation, polarization such that the tip of the electric field vector describes a helix. *Note 1:* The magnitude of the electric field vector is constant. *Note 2:* The projection of the tip of the electric field vector upon any fixed plane intersecting, and normal to, the direction of propagation, describes a circle. *Note 3:* A circularly polarized wave may be resolved into two linearly polarized waves in phase quadrature with their planes of polarization at right angles to each other. *Note 4:* Circular polarization may be referred to as *"right-hand"* or *"left-hand,"* depending on whether the helix describes the thread of a right-hand or left-hand screw, respectively.

Citizen Alert Program. Program in which residents of a community patrol streets and school areas with wireless phones and notify police of suspicious or illegal activity.

Cladding Diameter. In the cross section of a realizable optical fiber, ideally circular, but in practice assumed to a first approximation to be elliptical, the average of the diameters of the smallest circle that can be circumscribed about the cladding, and the largest circle that can be inscribed within the cladding.

Clamper. A diode circuit used to change the DC level of a waveform without distorting the waveform.

Clapp Oscillator. A variation of the Colpitts oscillator. An added capacitor is used to eliminate the effects of stray capacitance on the operation of the basic Colpitts oscillator.

Class A Amplifier. A linear amplifier biased so the active device conducts through 360 degrees of the input waveform.

Class B Amplifier. An amplifier with two active devices. The active components are biased so that each conducts for approximately 180 degrees of the input waveform cycle.

Class C Amplifier. An amplifier in which the active device conducts for less than 180 degrees of the input waveform cycle.

ClassLink. A program of the CTIA Foundation providing wireless phones to schools for teacher use and student Internet access

Clear Channel. 1. In radio broadcasting, a frequency assigned for the exclusive use of one entity. 2. In networking, a signal path that provides its full bandwidth for a user=s service. *Note:* No control or signaling is performed on this path.

Cleavage of Lateral Epitaxial Films for Transfer (CLEFT). A process for making inexpensive GaAs photovoltaic cells in which a thin film of GaAs is grown atop a thick, single-crystal GaAs (or other suitable material) substrate and then is cleaved from the substrate and incorporated into a cell, allowing the substrate to be reused to grow more thin-film GaAs.

Cleave. 1. In an optical fiber, a deliberate, controlled break, intended to create a perfectly flat endface, perpendicular to the longitudinal axis of the fiber. *Note:* A cleave is made by first introducing a microscopic fracture ("nick") into the fiber with a special tool, called a *"cleaving tool,"* which has a sharp blade of hard material, such as diamond, sapphire, or tungsten carbide. If proper tension is applied to the fiber as the nick is made, or immediately afterward (this may be done by the cleaving tool in some designs, or manually in other designs), the fracture will propagate in a controlled fashion, creating the desired endface. 2. To break a fiber in such a controlled fashion. *Note:* A good cleave is required for a successful splice of an optical fiber, whether by fusion or mechanical means. Also, some types of fiber-optic connectors do not employ abrasives and polishers. Instead, they use some type of cleaving technique to trim the fiber to its proper length, and produce a smooth, flat perpendicular endface.

CLEC (Competitive Local Exchange Carrier). A new entrant providing local wireline phone service.

Click-box Bidding. A method of bidding for FCC spectrum licenses in which the participant does not enter a bid amount but simply clicks on the computer screen to enter an amount shown. The method was first introduced in the auction of upper channel 800 MHz licenses in the fall of 1997.

Clipper. 1. A diode circuit used to eliminate part of a waveform. 2. A circuit or device that limits the instantaneous output signal amplitude to a predetermined maximum value, regardless of the amplitude of the input signal.

Clipping. Distortion caused by overdriving an amplifier.

Clock. A square waveform used for synchronizing and timing of several circuits.

Clone. A wireless phone programmed with stolen or duplicated electronic serial and mobile identification numbers.

Closed Circuit. 1. In radio and television transmission, pertaining to an arrangement in which programs are directly transmitted to specific users and not broadcast to the general public. 2. In telecommunications, a circuit dedicated to specific users. 3. A completed electrical circuit.

Closed-loop Gain. Gain of an amplifier when a feedback path is present.

Closed-loop Noise Bandwidth. The integral, over all frequencies, of the absolute value of the closed-loop transfer function of a phase-locked loop. *Note:* The closed-loop noise bandwidth, when multiplied by the noise spectral density, gives the output noise power in a phase-locked loop.

Closed-loop Transfer Function. A mathematical expression (algorithm) describing the net result of the effects of a closed (feedback) loop on the input signal to the circuits enclosed by the loop. *Note 1:* The closed-loop transfer function is measured at the output. *Note 2:* The output signal waveform can be calculated from the closed-loop transfer function and the input signal waveform.

Closed Waveguide. An electromagnetic waveguide (a) that is tubular, usually with a circular or rectangular cross section, (b) that has electrically conducting walls, (c) that may be hollow or filled with a dielectric material, (d) that can support a large number of discrete propagating modes, though only a few may be practical, (e) in which each discrete mode defines the propagation constant for that mode, (f) in which the field at any point is describable in terms of the supported modes, (g) in which there is no radiation field, and (h) in which discontinuities and bends cause mode conversion but not radiation.

CMOS. Subjective testing case: *Comparison Mean Opinion Score,* ITU-T P.800 subjective listening tests where test subjects compare samples of signals passing through a device under test with a known reference signal.

Semiconductor case: *Complimentary Metal Oxide Semiconductor,* a high speed technology used for integrated circuits and as CCDs, used for imaging in cameras, for example.

CMRS (Commercial Mobile Radio Service). An FCC designation for any carrier or licensee whose wireless network is connected to the public switched telephone network and/ or is operated for profit.

CNI. Calling Number Identification, in telephony.

C.O. Telco Central Office, where telecommunications equipment is located.

Coal. A black, solid fossil fuel found in the Earth. Coal is often burned to make electricity.

Coaxial Cable. 1. Transmission line in which the signal carrying conductor is covered by a dialectric and another conductor. 2. A cable consisting of a center conductor surrounded by an insulating material and a concentric outer conductor. *Note:* Coaxial cable is used primarily for wideband, video, or rf applications.

Co-branding. The use of two or more different brand names on a single product, such as wireless phones bearing the name and logo of both the manufacturer and wireless carrier.

Codec. COder - DECoder. A digital device for the encoding and decoding of video and/or audio signals, usually to permit their transmission in a compressed and/or encrypted form.

Code Conversion. 1. Conversion of signals, or groups of signals, in one code into corresponding signals, or groups of signals, in another code. 2. A process for converting a code of some predetermined bit structure, such as 5, 7, or 14 bits per character interval, to another code with the same or a different number of bits per character interval. *Note:* In code conversion, alphabetical order is not significant.

Coding. 1. In communications systems, the altering of the characteristics of a signal to make the signal more suitable for an intended application, such as optimizing the signal for transmission, improving transmission quality and fidelity, modifying the signal spectrum, increasing the information content, providing error detection and/or correction, and providing data security. *Note:* A single coding scheme usually does not provide more than one or two specific capabilities. Different codes have different advantages and disadvantages. 2. In communications and computer systems, implementing rules that are used to map the elements of one set onto the elements of another set, usually on a one-to-one basis. 3. The digital encoding of an analog signal and, conversely, decoding to an analog signal.

Coefficient of Coupling. The degree of coupling between two circuits.

Coercive Force (H). Magnetizing force needed to reduce residual magnetism in a material to zero.

Cogeneration. The process in which fuel is used to produce heat for a boiler-steam turbine or gas for a turbine. The turbine drives a generator that produces electricity, with the excess heat used for process steam.

Coherence Area. Pertaining to an electromagnetic wave, the area of a surface perpendicular to the direction of propagation, over which the electromagnetic wave maintains a specified degree of coherence. *Note:* The specified degree of coherence is usually taken to be 0.88 or greater.

Coherence Time. For an electromagnetic wave, the time over which a propagating wave may be considered coherent. *Note 1:* In long-distance transmission systems, the coherence time may be reduced by propagation factors such as dispersion, scattering, and diffraction. *Note 2:* In optical communications, coherence time, τ, is calculated by dividing the coherence length by the phase velocity of light in a medium; approximately given by $\tau = \lambda^2/(c\Delta\lambda)$ where λ is the central wavelength of the source, Δλ is the spectral width of the source, and c is the velocity of light in vacuum. *Note 3:* "*Coherence time*" is usually applied to the optical regime.

Collector. The semiconductor region in a bipolar junction transistor through which a flow of charge carriers leaves the base region.

Collector Characteristic Curve. A graph of collector voltage over collector current for a given base current.

Collocation. Placement of multiple antennas at a common physical site to reduce environmental impact and real estate costs and speed zoning approvals and net work deployment. Collocation can be affected by competitive and interference factors. Some companies act as brokers, arranging for sites and coordinating several carriers' antennas at a single site.

Color Burst. The part of an NTSC or PAL composite video signal which provides a frequency and phase reference for the demodulation of the color information in codecs and video monitors.

Color Code. Set of colors used to indicate value of a component.

Colpitts Oscillator. An oscillator with a pair of tapped capacitors in the feedback network.

COLT (Cell Site on Light Truck). A mobile site on a vehicle placed at a location to fill in or increase coverage.

Combined Collector. A photovoltaic device or module that provides useful heat energy in addition to electricity.

Common-anode Display. A multisegment light emitting diode (LED) with a single positive voltage input connection. Separate cathode connections are provided for each individual segment.

Common Battery. A single electrical power source used to energize more than one circuit, component, equipment, or system. *Note 1:* A common battery is usually an electrolytic device and is usually centrally located to the equipment that it serves. *Note 2:* In many telecommunications applications, the common battery is at a nominal B48 vdc. *Note 3:* A central office common battery supplies power to operate all directly connected instruments. *Note 4: Common battery* may include one or more power conversion devices to transform commercial power to direct current, with an electrolytic battery floating across the output.

Common-battery Signaling. Signaling in which the signaling power of a telephone is supplied by the serving switch. *Note:* In common-battery signaling, "talking power" may be supplied by common or local battery.

Common Carrier. A telephone company or similar supplier of non-private telecommunications services (telecommunications usage).

Common Cathode Display. A multisegment light emitting diode (LED) with a single negative voltage input connection. Separate anode connections are provided for each individual segment.

Common Base Amplifier. A BJT circuit in which the base connection is common to both input and output.

Common Collector Amplifier. A BJT circuit in which the collector connection is common to both input and output.

Common Drain Amplifier. A FET circuit in which the drain connection is common to both input and output.

Common Emitter Amplifier. A BJT circuit in which the emitter connection is common to both input and output.

Common Gate Amplifier. A FET circuit in which the gate connection is common to both input and output.

Common Source Amplifier. A FET circuit in which the source connection is common to both input and output.

Common-mode Rejection Ratio (CMRR). The ratio of op-amp differential gain to common-mode gain. A measure of an op-amp's ability to reject common-mode signals such as noise.

Common-mode Signals. Signals that appear simultaneously at two inputs of an operational amplifier (op-amp). Common mode signals are always equal in amplitude and phase.

Communications-electronics (C-E). The specialized field concerned with the use of electronic devices and systems for the acquisition or acceptance, processing, storage, display, analysis, protection, disposition, and transfer of information. *Note:* C-E includes the wide range of responsibilities and actions relating to (a) electronic devices and systems used in the transfer of ideas and perceptions, (b) electronic sensors and sensory systems used in the acquisition of information devoid of semantic influence, and (c) electronic devices and systems intended to allow friendly forces to operate in hostile environments and to deny to hostile forces the effective use of electromagnetic resources.

Communications Jamming (COMJAM). 1. The portion of electronic jamming that is directed against communications circuits and systems. 2. The prevention of successful radio communications by the use of electromagnetic signals, *i.e.*, the deliberate radiation, reradiation, or reflection of electromagnetic energy with the objective of impairing the effective use of electronic communications systems. *Note:* The aim of communications jamming is to prevent communications by electromagnetic means, or at least to degrade communications sufficiently to cause delays in transmission and reception. Jamming may be used in conjunction with deception to achieve an overall electronic-countermeasure (ECM) plan implementation.

Compact Fluorescent Lights. Lights that use a lot less energy than regular light bulbs. We can use compact fluorescent lights for reading lights and ceiling lights.

Compander. COMPresser/expANDER. Both A-law and ì-law companders are used in telephony to improve the subjective signal-to-(quantizing)-noise ratio for audio communications and telephony.

Comparitor. An op-amp circuit that compares two inputs and provides a DC output indicating the polarity relationship between the inputs.

Complementary Symmetry Amplifier. A class B amplifier using matched complementary transistors. Does not require a phase inverter for push-pull output.

Complementary Transistors. Two transistors, one NPN and one PNP having near identical characteristics. N-channel and P-channel FETs can also be complementary.

Complex Numbers. Numbers composed of a real number part and an imaginary number part.

Compliance. The maximum possible peak-to-peak output of an amplifier.

Component. 1. An assembly, or part thereof, that is essential to the operation of some larger assembly and is an immediate subdivision of the assembly to which it belongs. *Note:* For example, a radio receiver may be a component of a complete radio set consisting of a combined transmitter-receiver, *i.e.*, a transceiver. The same radio receiver could also be a subsystem of the combined transmitter-receiver, in which case the IF amplifier section would be a component of the receiver but not of the radio set. Similarly, within the IF amplifier section, items, such as resistors, capacitors, vacuum tubes, and transistors, are components of that section. 2. In logistics, a part, or combination of parts having a specified function, that can only be installed or replaced as an entity. 3. In material, an assembly or any combination of parts, subassemblies, and assemblies mounted together in manufacture, assembly, maintenance, or rebuild. 4. Semiconductors and smaller elements used in handsets and infrastructure equipment.

Component Video Signal. A video signal where the red (R), green (G) and blue (B) picture components are present as individual signals. Synchronization information may be included with the G signal or may be a separate signal. Less commonly (for videoconferencing, s-video), luminance and chrominance signals (YIQ or YUV) may be provided as separate signals, also referred to as component signals.

Composite Signaling (CX). Signaling in which an arrangement is made to provide direct current signaling and dial pulsing beyond the range of conventional loop signaling. *Note:* Composite signaling, like DX signaling, permits duplex operation, *i.e.,* permits simultaneous two-way signaling. *Synonym* CX signaling.

Compound Signal. In ac signaling, a signal consisting of the simultaneous transmission of more than one frequency. *Note:* An example of compound signaling is dual-tone multifrequency (DTMF) signaling.

Compressed-air Energy Storage (CAES). CAES plants use off-peak electrical energy to compress air into underground storage reservoirs for storage until times of peak or intermediate electricity demand. Wind power offers a good opportunity for charging CAES storage. The storage is typically underground in natural aquifers, depleted oil or gas fields, mined salt caverns, or excavated or natural rock caverns. To generate power, the compressed air is first heated by gas burners, then passed through a turbine.

Composite Video Signal. A video signal incorporating luminance, chrominance and synchronizing information.

Compression, Data. The process of reducing the quantity of data necessary to transmit or store. See below:

Compression, Lossless : A process allowing data compression and its subsequent restoration to its original form without any loss of information. PKZip, ZOO, RAR and ARC are common examples. See also *Huffman Coding.*

Compression, Lossy : A data compression process that makes subsequent complete recovery of the original data impossible. This approach is commonly used for still- or motion-images where the recovered image must only be subjectively acceptable. Algorithm examples include JPEG, MPEG1, MPEG2, MPEG4, H.261 and H.263. Note that lossy compression can provide significantly greater compression than lossless compression techniques. See also *Cosine Transform*.

Compression Ratio. 1. In signal compression, the ratio of the dynamic range of compressor input signals to the dynamic range of the compressor output signals. *Note:* The compression ratio is usually expressed in dB. For example, a 40-dB input range compressed to a 30-dB output range would be equivalent to a 10-dB compression. 2. In digital facsimile, the ratio of the total pels scanned for the object to the total encoded bits sent for picture information. 3. The ratio of the gain of a device at a low power level to the gain at some higher level. *Note:* The compression ratio is usually expressed in dB.

Compressor. A nonlinear analog device that has a lower gain at higher input levels than at lower input levels. *Note:* A compressor is used to allow signals with a large dynamic range to be sent through devices or circuits with a smaller dynamic range .

Concentrator. A PV module that uses optical elements to increase the amount of sunlight incident on a PV cell. Concentrating arrays must track the sun and use only the direct sunlight because the diffuse portion cannot be focused onto the PV cells.

Concentrator (module, array, or collector). An arrangement of photovoltaic cells that includes a lens to concentrate sunlight onto small-area cells. Concentrators can increase the power flux of sunlight hundreds of times.

Concentricity Error. In an optical fiber, the distance between the center of the two concentric circles that specify the cladding diameter and the center of the two concentric circles that specify the core diameter. *Note:* The concentricity error is used in conjunction with tolerance fields to specify or characterize optical fiber core and cladding geometry. *Synonyms* core eccentricity, core-to-cladding concentricity, core-to-cladding eccentricity, core-to-cladding offset.

Conditional Replenishment. In motion video compression, referrs to the conditional transmission of information from blocks of pixels in adjacent video frames only if the content has changed materially or after a certain period of time has elapsed since the last refresh.

Conditioned Diphase Modulation. A form of diphase modulation, combined with signal conditioning, that (a) eliminates the dc component of the signal, (b) enhances timing recovery, and (c) facilitates transmission over voice frequency (VF) circuits or coaxial cables.

Conditioning Equipment. 1. At junctions of circuits, equipment used to obtain desired circuit characteristics, such as matched transmission levels, matched impedances, and equalization between facilities. 2. Corrective networks used to improve data transmission, such as equalization of the insertion-loss-vs.-frequency characteristic and the envelope delay distortion over a desired frequency range.

Conduction Band. 1. In a semiconductor, the range of electron energy, higher than that of the valence band, sufficient to make the electrons free to move from atom to atom under the influence of an applied electric field and thus constitute an electric current. 2. In the atomic structure of a material, a partially filled or empty energy level in which electrons are free to move, thus allowing the material to conduct an electrical current upon application of an electric field by means of an applied voltage.

Conduction Level. Energy level at which electrons are not bound to (orbiting) a specific atomic nucleus but are free to wander among the atoms.

Conductive Coupling. Energy transfer achieved by means of physical contact, *i.e.*, coupling other than inductive or capacitive coupling. *Note 1:* Conductive coupling may be achieved by wire, resistor, or common terminal, such as a binding post or metallic bonding. *Note 2:* Conductive coupling passes the full spectrum of frequencies, including dc. *Synonym* direct coupling.

Connection. 1. A provision for a signal to propagate from one point to another, such as from one circuit, line, subassembly, or component to another. 2. An association established between functional units for conveying information.

Constant Current Circuit. Circuit used to maintain constant current to a load having resistance that changes.

Constant-speed Wind Turbines. Turbines that operate at a constant rotor revolutions per minute (RPM) and are optimized for energy capture at a given rotor diameter at a particular speed in the wind power curve.

Contact. Current carrying part of a switch, relay or connector.

Contact Resistance. The resistance between metallic contacts and the semiconductor.

Content Services. Paging service, beyond telephone number alerts, that include news and sports headlines, personalized stock quotes, driving directions, restaurant reviews and information contained on Internet sites.

Continuity. Occurs when a complete path for current exists.

Continuous Presence. A videoconferencing technique, allowing all participants to see all other perticipants, all the time. It makes use of a video processing, transmission and display technique generally implemented by extracting the horizontal center half of video images from two cameras, and electronically stacking the two center halves into a single video signal or data stream for transmission. At the receive location(s), the images may be displayed one above the other on a single video monitor or alternatively, viewed on two side-by-side video monitors. Continuous presence can be used with video codecs.

Control Channel. A logic channel carrying network information rather than the actual voice or data messages transmitted over the network.

Control of Electromagnetic Radiation. 1. Measures taken to minimize unintended electromagnetic radiation emanating from a system or component and to minimize electromagnetic interference. *Note:* Control of electromagnetic radiation is exercised for purposes of security and the reduction of interference, especially on ships and aircraft. 2. A national operational plan to minimize the use of electromagnetic radiation in the United States and its possessions and the Panama Canal Zone in the event of attack or imminent threat thereof, as an aid to the navigation of hostile aircraft, guided missiles, or other devices.

Conventional Current Flow. Concept of current produced by the movement of positive charges towards the negative terminal of a source.

Conversion Efficiency (cell or module). The ratio of the electric energy produced by a photovoltaic device (under one-sun conditions) to the energy from sunlight incident upon the cell.

Copper Indium Diselenide ($CuInSe_2$, or CIS). A polycrystalline thin-film photovoltaic material (sometimes incorporating gallium (CIGS) and/or sulfur).

Copper Loss. Power lost in transformers, generators, connecting wires and other parts of a circuit due to current flow through the resistance of copper conductors.

Cord Circuit. A switchboard circuit in which a plug-terminated cord is used to establish connections manually between user lines or between trunks and user lines. *Note:* A number of cord circuits are furnished as part of the switchboard position equipment. The cords may be referred to as front cord and rear cord or trunk cord and station cord. In modern cordless switchboards, the cord-circuit function is switch operated and may be programmable.

Core. 1. The central region about the longitudinal axis of an optical fiber, which region supports guiding of the optical signal. *Note 1:* For the fiber to guide the optical signal, the refractive index of the core must be slightly higher than that of the cladding. *Note 2:* In different types of fibers, the core and core-cladding boundary function slightly differently in guiding the signal. Especially in single-mode fibers, a significant fraction of the energy in the bound mode travels in the cladding. 2. A piece of ferromagnetic material, usually toroidal in shape, used as a component in a computer memory device. *Note:* The type of memory referred to has very limited application in today=s computer environment. It has been largely replaced by semiconductor and other technologies. 3. The material at the center of an electromechanical relay or solenoid, about which the coil is wound.

Cosine Transform. The two-dimensional discrete cosine transform (DCT) developed in 1981 by Wen-hsiung Chen (then Chief Scientist, Compression Labs, Inc., San Jose, CA). DCT is the basis of most lossy still-image and motion video compression algorithms including JPEG, MPEG1, MPEG2, MPEG4, H.261 and H.263 standards. The DCT converts blocks of picture elements from the vertical and horizontal spatial domains to the transform domain. Put another way, blocks of pixels in rows and columns are converted to the frequency domain. Depending on a number of factors, the coefficients produced may be truncated by varying degrees prior to storage or transmission without subjectively affecting the recovered images.

Cosmic Noise. Random noise that originates outside the Earth=s atmosphere. *Note:* Cosmic noise characteristics are similar to those of thermal noise. Cosmic noise is experienced at frequencies above about 15 MHz when highly directional antennas are pointed toward the Sun or to certain other regions of the sky such as the center of the Milky Way Galaxy. *Synonym* galactic radio noise.

Cost Recovery. Reimbursement to CMRS providers of both recurring and nonrecurring costs associated with any services, operation, administration or maintenance of wireless E911 service. Costs include, but are not limited to, the costs of design, development, upgrades, equipment, software and other expenses associated with the implementation of wireless E911 service.

Coulomb. Unit of electric charge. A negative coulomb charge consists of 6.24×10^{18} electrons.

Counter Electromotive Force (counter emf). Voltage induced into an inductor due to an alternating or pulsating current. Counter emf is always in polarity opposite to that of the applied voltage. Opposing a change of current.

Coupled Modes. 1. In fiber optics, a mode that shares energy among one or more other modes, all of which propagate together. *Note:* The distribution of energy among the coupled modes changes with propagation distance. 2. In microwave transmission, a condition where energy is transferred from the fundamental mode to higher order modes. *Note:* Energy transferred to coupled modes is undesirable in usual microwave transmission in a waveguide.

Coupling. To electronically connect two circuits so that signal will pass from one to the other.

Coupling Coefficient. A number that expresses the degree of electrical coupling that exists between two circuits. *Note:* The coupling coefficient is calculated as the ratio of the mutual impedance to the square root of the product of the self-impedances of the coupled circuits, all impedances being expressed in the same units.

Coupling Efficiency. In fiber optics, the efficiency of optical power transfer between two optical components. *Note 1:* The transfer may take place (a) between an active component, such as an LED, and a passive component, such as an optical fiber, or (b) between two passive components such as two optical fibers. *Note 2:* Coupling efficiency is usually expressed as the ratio, converted to percent, of the input power, *i.e.,* the available power from one component, to the power transferred to the other component.

Covalent Bond. The way some atoms complete their valence shells by sharing valence electrons with neighbouring atoms.

Covered SMR. A subset of specialized mobile radio operators subject to a particular set of regulations. A definition developed in December 1997 during the implementation of E911 regulations encompasses operators whose networks use intelligent switching capabilities and offer seamless hand-off to customers. An earlier, broader definition, still applicable in other contexts, encompasses operators that provided two-way, real-time voice services across geographic license areas in the 800 MHz and 900 MHz bands.

COW (Cell Site on Wheels). A mobile site placed at a location to fill in or increase coverage.

CPE (Consumer Premise Equipment). Telephones, PBXs and other communications devices located in the home or office.

CPNI (Customer Proprietary Network Information). The carrier's data about a specific customer's service and usage. The FCC restricts CPNI use in marketing, banning win-back efforts specifically aimed at high-usage customers who have quit a network.

Cracker. Slang term for someone who breaks an encrypted computer code or circuitry.

CRC. Cyclic Redundancy Code. An error checking method for digital signals.

Critical Frequency. 1. In radio propagation by way of the ionosphere, the limiting frequency at or below which a wave component is reflected by, and above which it penetrates through, an ionospheric layer. 2. At vertical incidence, the limiting frequency at or below which incidence, the wave component is reflected by, and above which it penetrates through, an ionospheric layer. *Note:* The existence of the critical frequency is the result of electron limitation, *i.e.,* the inadequacy of the existing number of free electrons to support reflection at higher frequencies.

Crossover Distortion. Distortion caused by both devices in a class B amplifier being cut-off at the same time.

Crosstalk. Interference in a wireless communications system that stems from other conversations in nearby cells using the same channel.

Crowbar. Circuit used to protect the output of a source from a short circuited load. Load current is limited to a value the source can deliver without damage.

CRT. Abbreviation for cathode ray tube.

Cryptography. Originally, the study of secret writing, including codes and ciphers. Nowadays, the science of mathematical techniques called encryption (which see) allowing distribution or transmission of information or data so that it can only be read or used by the intended recipients.

Crystal. Natural or synthetic piezoelectric or semiconductor material with atoms arranged with some degree of geometric regularity.

Crystal-controlled Oscillator. Oscillator that uses a quartz crystal in its feedback path to maintain a stable output frequency.

Crystal Oscillator (XO). An oscillator in which the frequency is controlled by a piezoelectric crystal. *Note 1:* A crystal oscillator may require controlled temperature because its operating frequency is a function of temperature. *Note 2:* Types of crystal oscillators include voltage-controlled crystal oscillators (VCXO), temperature-compensated crystal oscillators (TCXO), oven-controlled crystal oscillators (OCXO), temperature-compensated-voltage controlled crystal oscillators (TCVCXO), oven-controlled voltage-controlled crystal oscillators (OCVCXO), microcomputer-compensated crystal oscillators (MCXO), and rubidium crystal oscillators (RbXO).

CSDN. Circuit Switched Digital Network.

CSU. Channel Service Unit. Usually a customer-owned CPE device that provides an interface between digital equipment (like codecs) and transmission facilities that comply with FCC or other regulatory requirements. A CSU often includes network switching and control and/or line-conditioning capabilities.

CTI. *Computer Telephony Integration*, computer applications which augment telephony.

CTIA (Cellular Telecommunications Industry Association). A trade group representing cellular, PCS and enhanced specialized mobile radio carriers.

Current. Measured in amperes, it is the flow of electrons through a conductor. Also know as electron flow.

Current Amplifier. Amplifier to increase signal current.

Current at Maximum Power (Imp). The current at which maximum power is available from a module. [UL 1703]

Current Divider. Parallel network designed to divide the total current of a circuit

Current Feedback. Feedback configuration where a portion of the output current is fed back to the amplifier input.

Current-limiting Resistor. Resistor in the path of current flow to control the amount of current drawn by a device.

Current Mirror. Term used to describe the fact that DC current through the base circuit of a class B amplifier is approximately equal to the DC collector current.

Customer Acquisition Cost. The average cost to a carrier of signing up an individual subscriber. Some of the factors included in the cost are handset subsidies, marketing, advertising and promotions.

Cutoff. Condition when an active device is biased such that output current is near zero or beyond zero.

Cutoff Frequency. 1. The frequency either above which or below which the output of a circuit, such as a line, amplifier, or filter, is reduced to a specified level. 2. The frequency below which a radio wave fails to penetrate a layer of the ionosphere at the incidence angle required for transmission between two specified points by reflection from the layer.

Cutoff Wavelength. 1. The wavelength corresponding to the cutoff frequency. 2. In an uncabled single-mode optical fiber, the wavelength greater than which a particular waveguide mode ceases to be a bound mode. *Note 1:* The cutoff wavelength is usually taken to be the wavelength at which the normalized frequency is equal to 2.405. *Note 2:* The *cabled* cutoff wavelength is usually considered to be a more functional parameter because it takes into consideration the effects of cabling the fiber.

CVSDM. Continuously Variable Slope Delta Modulation. A digital audio coding technique providing good quality audio, especially robust under high bit error rate conditions.

Cycle. When a repeating wave rises from zero to a positive maximum then back to zero and on to a negative maximum and back to zero it is said to have completed one cycle.

Cycle Life. Number of discharge-charge cycles that a battery can tolerate under specified conditions before it fails to meet specified criteria as to performance (e.g., capacity decreases to 80-percent of the nominal capacity).

Czochralski Process. A method of growing large size, high quality semiconductor crystal by slowly lifting a seed crystal from a molten bath of the material under careful cooling conditions.

D

D4. A T1-transmission framing and synchronization method, consisting of a series of digital 1's and 0's in a unique order, each member of the series being sequentially inserted in a T1 stream after every 193 data bits. See also *B8ZS*.

DAB. *Digital Audio Broadcast*, digital audio signal added to a conventional FM broadcast signal following FM modulation. Compatible with standard FM radio broadcast receivers, a special decoder is needed to hear the DAB signal.

DAC. Abbreviation for "digital to analog converter."

DACS. Digital Access Cross-connect System (which see). Included as part of public and private conventional digital networks, a DACS may commonly switch DS3, DS1 and DS0 and other high level digital signals and is useful for configuring complex networks. A DACS is not a substitute for an MCU.

Damping. Reduction in magnitude of oscillation due to energy being dissipated as heat.

D-AMPS (Digital AMPS). Used by Ericsson Inc. to describe IS-136 time division multiple access technology.

Dangling Bonds. A chemical bond associated with an atom on the surface layer of a crystal. The bond does not join with another atom of the crystal, but extends in the direction of exterior of the surface.

Darlington Pair. An amplifier consisting of two bipolar junction transistors with their collectors connected together and the emitter of one connected to the base of the other. Circuit has an extremely high current gain and input impedance.

Data Compaction. The reduction of the number of data elements, bandwidth, cost, and time for the generation, transmission, and storage of data without loss of information by eliminating unnecessary redundancy, removing irrelevancy, or using special coding. *Note 1:* Examples of data compaction methods are the use of fixed-tolerance bands, variable-tolerance bands, slope-keypoints, sample changes, curve patterns, curve fitting, variable-precision coding, frequency analysis, and probability analysis. *Note 2:* Simply squeezing noncompacted data into a smaller space, for example by increasing packing density or by transferring data on punched cards onto magnetic tape, is not data compaction. *Note 3:* Whereas data compaction reduces the amount of data used to represent a given amount of information, data compression does not.

Data Transmission Circuit. The transmission media and the intervening equipment used for the transfer of data between data terminal equipments (DTEs). *Note 1:* A data transmission circuit includes any required signal conversion equipment. *Note 2:* A data transmission circuit may transfer information in (a) one direction only, (b) either direction but one way at a time, or (c) both directions simultaneously.

dB. Decibel. One-tenth of a Bell, a logarithmic measurement of acoustic, electromagnetic or electrical power, voltage or current ratios.

dBa. *Abbreviation for* decibels adjusted. Weighted absolute noise power, calculated in dB referenced to 3.16 picowatts (B85 dBm), which is 0 dBa. *Note:* The use of F1A-line or HA1-receiver weighting must be indicated in parentheses as required. A one-milliwatt, 1000-Hz tone will read +85 dBa, but the same power as white noise, randomly distributed over a 3-kHz band (nominally 300 to 3300 Hz), will read +82 dBa, due to the frequency weighting. *Synonym* dBrn adjusted.

dBa(F1A). Weighted absolute noise power in dBa, measured by a noise measuring set with F1A-line weighting. *Note:* F1A weighting is no longer used for DOD applications.

dBa(HA1). Weighted noise power in dBa, measured across the receiver of a 302-type or similar subscriber set, by a noise measuring set with HA1-receiver weighting. *Note:* HA1 weighting is no longer used in DOD applications.

dBa0. Noise power in dBa referenced to or measured at zero transmission level point (0TLP), also called a point of zero relative transmission level (0 dBr). *Note:* It is preferred to convert noise readings from dBa to dBa0, as this makes it unnecessary to know or state the relative transmission level at the point of actual measurement.

dBc. *Abbreviation.* dB relative to the carrier power.

dBi. *Abbreviation.* In the expression of antenna gain, the number of decibels of gain of an antenna referenced to the zero dB gain of a free-space isotropic radiator.

dBrn. *Abbreviation.* dB above reference noise. *Note 1:* Weighted noise power in dB is referred to 1.0 picowatt. Thus, 0dBrn = B90 dBm. Use of 144-line, 144-receiver, or C-message weighting, or flat weighting, must be indicated in parentheses as required. *Note 2:* With C-message weighting, a one-milliwatt, 1000-Hz tone will read +90 dBrn, but the same power as white noise, randomly distributed over a 3-kHz band will read approximately +88.5 dBrn (rounded off to +88 dBrn), because of the frequency weighting. *Note 3:* With 144 weightings, a one-milliwatt, 1000-Hz white noise tone will also read +90 dBrn, but the same 3-kHz power will only read +82 dBrn, because of the different frequency weighting.

DBS. *Direct Broadcast Satellite*, C- and Ku-band satellite services providing television, sound, video and data signals directly to home users. The signals are often encrypted, intended for reception by subscribers, requiring an IRD for reception.

DC. Abbreviation for "direct current".

DCE. *Abbreviation for* data circuit-terminating equipment. 1. In a data station, the equipment that performs functions, such as signal conversion and coding, at the network end of the line between the data terminal equipment (DTE) and the line, and that may be a separate or an integral part of the DTE or of intermediate equipment. 2. The interfacing equipment that may be required to couple the data terminal equipment (DTE) into a transmission circuit or channel and from a transmission circuit or channel into the DTE. *Synonyms* data communications equipment *(deprecated)*, data set *(deprecated)*. 3. A configuration protocol identifying Data Communication Equipment, the type of data connection at equipment that connects to a communications facility. This consideration is important when interconnecting RS-449, RS-232, V.35, X.21 and other similar digital communications signals. Compare *DTE. Distributed Computing Environment*, a complex middleware platform providing security services.

D Channel. In ISDN, the 16-kb/s segment of a 144-kb/s, full-duplex subscriber service channel that is subdivided into 2B+D channels, *i.e.*, into two 64-kb/s clear channels and one 16-kb/s channel for the ISDN basic rate. *Note 1:* The D channel is usually used for out-of-band signaling. The two 64-kb/s clear channels are used for subscriber voice and data services. *Note 2:* The D-channel specifications are addressed in the CCITT Recommendation for the Integrated Services Digital Network (ISDN). *Note 3:* The D-channel may be 64 kb/s for the primary rate ISDN service.

DC Load Line. A graph representing all possible combinations of voltage and current for a given load resistor in an amplifier.

DCMA (Dynamic Channel Multicarrier Architecture). A technology developed by ComSpace Corp. used for specialized mobile radio networks that can configure the number and bandwidth of voice and data channels based on a carrier's requirements.

DC Offset. The change in input voltage required to produce a zero output voltage when no signal is applied to an amplifier.

DC Power Supply. Any source of DC power for electrical equipment.

DCR. *Degradation Category Rating*, an ITU-T P.800 subjective testing protocol, similar to CCR but is based on the test subject always listening first to the reference signal followed by the signal passed through the device under test, a test performed by HEI.

DCS 1800 (Digital Cellular System). A global system for mobile communications-based PCS network used outside of the U.S.

DCT (Discrete Cosine Transform). see *Cosine Transform.*

DC to DC Converter. Electronic circuit to convert dc voltages (e.g., PV module voltage) into other levels (e.g., load voltage). Can be part of a maximum power point tracker (MPPT).

Dead Short. Short circuit having zero resistance.

Dead Spot. An area within a wireless communications system where there is no coverage. This can occur because an cell site cannot be located close enough to the area without coverage, or because two signals on the same channel interfere with each other, causing them to "phase out" the other signa

Decade. A frequency factor of ten.

Deception Repeater. A device that can (a) receive a signal, (b) amplify, delay, or otherwise manipulate the signal, and (c) retransmit it solely for creating deception.

Decibel (dB). a logarithmic representation of gain or loss.

Decision Circuit. A circuit that measures the probable value of a signal element and makes an output signal decision based on the value of the input signal and a predetermined criterion or criteria.

DECT (Digital European Cordless Telephone). 1. Used as a cordless phone and wireless office phone system in Europe. 2. *Digitally Enhanced Cordless Telephone*, operates in the 1880-GHz - 1900-GHz band using Bluetooth's GFSK (MC/TDMA/TDD) technology.

Deep Discharge. Discharging a battery to 20-percent or less of its full charge.

Deeply Depressed Cladding Fiber. An optical fiber construction, usually a single-mode fiber, that has an outer cladding of approximately the same refractive index as the core, and an inner cladding of very low (depressed) refractive index material between them.

Deflection. 1. A change in the direction of a traveling particle, usually without loss of particle kinetic energy, representing a change in velocity without a change in the scalar speed of the particle. 2. A change in the direction of a wave, beam, electron, or other entity, such as might be accomplished by an electric or magnetic field. *Note:* If the deflection is caused by a prism (refraction), a mirror (reflection), or optical grating (diffraction), the specific applicable term should be used.

Degenerative Feedback. Also called negative feedback. A portion of the output of an amplifier is inverted and connected back to the input. This controls the gain of the amplifier and reduces distortion and noise.

Degree of Start-stop Distortion. 1. In asynchronous data transmission, the ratio of (a) the absolute value of the maximum measured difference between the actual and theoretical intervals separating any significant instant of modulation (or demodulation) from the significant instant of the start element immediately preceding it to (b) the unit interval. 2. The highest absolute value of individual distortion affecting the significant instants of a start-stop modulation. *Note:* The degree of distortion of a start-stop modulation (or demodulation) is usually expressed as a percentage. Distinction can be made between the degree of late (positive) distortion and the degree of early (negative) distortion.

Dejitterizer. A device that reduces jitter in a digital signal. *Note 1:* A dejitterizer usually consists of an elastic buffer in which the signal is temporarily stored and then retransmitted at a rate based on the average rate of the incoming signal. *Note 2:* A dejitterizer is usually ineffective in dealing with low-frequency jitter, such as waiting-time jitter.

Delay Distortion. In a waveform consisting of two or more wave components at different frequencies, distortion caused by the difference in arrival times of the frequency components at the output of a transmission system. *Synonyms* time-delay distortion.

Delay Line. 1. A transmission line, or equivalent device, used to delay a signal. 2. A single-input-channel device, such as a single-input sequential logic element, in which the output channel state at a given instant, *t*, is the same as the input channel state at the instant *tBn*, where n is a number of time units, *i.e.*, the input sequence undergoes a delay of *n* time units, such as *n* femtoseconds, nanoseconds, or microseconds. *Note:* The delay line may have additional taps yielding output channels with values less than *n*.

Delay Time. The time for collector current to reach 10% of its maximum value in a BJT switching circuit.

Delta Modulation (DM). Analog-to-digital signal conversion in which (a) the analog signal is approximated with a series of segments, (b) each segment of the approximated signal is compared to the original analog wave to determine the increase or decrease in relative amplitude, (c) the decision process for establishing the state of successive bits is determined by this comparison, and (d) only the change of information is sent, *i.e.*, only an increase or decrease of the signal amplitude from the previous sample is sent whereas a no-change condition causes the modulated signal to remain at the same 0 or 1 state of the previous sample. *Note:* Examples of delta modulation are continuously variable slope delta modulation, delta-sigma modulation, and differential modulation.

Demand Factor. 1. The ratio of (a) the maximum real power consumed by a system to (b) the maximum real power that would be consumed if the entire load connected to the system were to be activated at the same time. *Note:* The maximum real power is usually integrated over a specified time interval, such as 15 or 30 minutes, and is usually expressed in kilowatts. The real power that would be consumed if the entire load connected to the system were to be activated at the same time is obtained by summing the power required by all the connected equipment. This load is expressed in kilowatts if the consumed real power is expressed in kilowatts. 2. The ratio of (a) the maximum power, integrated over a specified time interval, such as 15 or 30 minutes, and usually expressed in kilowatts, consumed by a system, to (b) the maximum volt-amperes, expressed in kilovolt-amperes if the power is expressed in kilowatts, integrated over a time interval of the same duration, though not necessarily during the same interval. *Note:* Charges for electrical power may be based on the demand factor as well as the kilowatt-hours of electrical energy consumed.

Demand Load. 1. In general, the total power required by a facility. *Note:* The demand load is the sum of the operational load (including any tactical load) and nonoperational demand loads. It is determined by applying the proper demand factor to each of the connected loads and a diversity factor to the sum total. 2. At a communications center, the power required by all automatic switching, synchronous, and terminal equipment (operated simultaneously on-line or in standby), control and keying equipment, plus lighting, ventilation, and air- conditioning equipment required to maintain full continuity of communications. 3. The power required for ventilating equipment, shop lighting, and other support items that may be operated simultaneously with the technical load. 4. The sum of the technical demand and nontechnical demand loads of an operating facility. 5. At a receiver facility, the power required for all receivers and auxiliary equipment that may be operated on prime or spare antennas simultaneously, those in standby condition, multicouplers, control and keying equipment, plus lighting, ventilation, and air conditioning equipment required for full continuity of communications. 6. At a transmitter facility, the power required for all transmitters and auxiliary equipment that may be operated on prime or spare antennas or dummy loads simultaneously, those in standby condition, control and keying equipment, plus lighting, ventilation, and air conditioning equipment required for full continuity of communications.

De Minimus. Having such minimal impact as to not influence the matter at hand. Costs of providing a service, for instance, may be so low that the party requesting the service sees them as de minimus and an invalid reason for not offering a service.

DEMS (Digital Electronic Message Service). The FCC wants to relocate this service from 18 GHz to 24 GHz.

Dendrite. A slender threadlike spike of pure crystalline material, such as silicon.

Dendritic Web Technique. A method for making sheets of polycrystalline silicon in which silicon dendrites are slowly withdrawn from a melt of silicon whereupon a web of silicon forms between the dendrites and solidifies as it rises from the melt and cools.

Depletion Region. The area surrounding a pn junction that is depleted of carriers.

Depletion Mode. In a FET, an operating mode where reverse gate-source voltage is used to deplete the channel of free carriers. This reduces the size of the channel and increases its resistance.

Depletion-mode MOSFET. A MOSFET designed to operate in either depletion mode or enhancement mode.

Depletion Zone. Same as cell barrier. The term derives from the fact that this microscopically thin region is depleted of charge carriers (free electrons and holes).

DES. Data Encryption Standard. A 64-bit private-key encryption technique developed by the US *NBS*. An encryption system that encodes communications, providing security and "privacy" of communication by making interception and decoding impossible (for most practical purposes). Double DES or DES2 uses tandem DES algorithms for greater security. Variants may use variable-length keys.

Design Objective (DO). In communications systems, a desired performance characteristic for communications circuits and equipment that is based on engineering analyses, but (a) is not considered feasible to mandate in a standard, or (b) has not been tested. *Note 1:* DOs are used because applicable systems standards are not in existence. *Note 2:* Examples of reasons for designating a performance characteristic as a DO rather than as a standard are (a) it may be bordering on an advancement in the state of the art, (b) the requirement may not have been fully confirmed by measurement or experience with operating circuits, and (c) it may not have been demonstrated that the requirement can be met considering other constraints, such as cost and size. *Note 3:* A DO is sometimes established in a standard for developmental consideration. A DO may also specify a performance characteristic that may be used in the preparation of specifications for development or procurement of new equipment or systems.

Detection. 1. The recovery of information from an electrical or electromagnetic signal. *Note:* Conventional radio waves are usually detected by heterodyning, *i.e.*, coherent reception/detection. In this method of reception/detection, the received signal is mixed, in some type of nonlinear device, with a signal from a local oscillator, to produce an intermediate frequency, *i.e.*, beat frequency, from which the modulating signal is recovered, *i.e.*, detected. The inherent instabilities of available optical sources have, until relatively recently, prevented practical use of coherent reception/detection in optical communication receivers. At present, coherent optical receivers, using sophisticated technology, are just beginning to emerge from the laboratory into the field. Virtually all existing optical receivers employ direct detection; that is, the received optical signal impinges directly onto a detector. Direct detection is less sensitive than coherent detection. 2. In tactical operations, the perception of an object of possible military interest but unconfirmed by recognition. 3. In surveillance, the determination and transmission by a surveillance system that an event has occurred.

Device. A component or part.

Diac. A two terminal bidirectional thyristor. Has a symmetrical switching mode.

Dichroic Filter. An optical filter that reflects one or more optical bands or wavelengths and transmits others, while maintaining a nearly zero coefficient of absorption for all wavelengths of interest. *Note:* A dichroic filter may be high-pass, low-pass, band-pass, or band rejection.

Dielectric. 1. A substance in which an electric field may be maintained with zero or near-zero power dissipation, *i.e.*, the electrical conductivity is zero or near zero. *Note 1:* A dielectric material is an electrical insulator. *Note 2:* In a dielectric, electrons are bound to atoms and molecules, hence there are few free electrons. 2. Pertaining to a substance that has a zero or near zero electrical conductivity.

Dialectric Constant. Property of a material that determines how much electrostatic energy can be stored per unit volume when unit voltage is applied.

Dielectric Lens. In the radio regime, a lens made of dielectric material that refracts radio waves in the same manner that an optical lens refracts light waves.

Dielectric Strength. 1. Of an insulating material, the maximum electric field strength that it can withstand intrinsically without breaking down, *i.e.*, without experiencing failure of its insulating properties. *Note:* The theoretical dielectric strength of a material is an intrinsic property of the bulk material and is not dependent on the configuration of the material or the electrodes with which the field is applied. 2. For a given configuration of dielectric material and electrodes, the minimum electric field that produces breakdown. *Note 1:* At breakdown, the electric field frees bound electrons, turning the material into a conductor. *Note 2:* The field strength at which breakdown occurs in a given case is dependent on the respective geometries of the dielectric (insulator) and the electrodes with which the electric field is applied, as well as the rate of increase at which the electric field is applied. *Note 3:* The electric field strength is usually expressed in volts per meter.

Dielectric Waveguide. A waveguide that consists of a dielectric material surrounded by another dielectric material, such as air, glass, or plastic, with a lower refractive index. *Note 1:* An example of a dielectric waveguide is an optical fiber. *Note 2:* A metallic waveguide filled with a dielectric material is not a dielectric waveguide.

Differential Amplifier. An amplifier in which the output is in proportion to the differences between voltages applied to its two inputs.

Differential Interframe Coding. In motion video compression, the transmission of only the difference information between adjacent video frames.

Differentiator. A circuit in which the output voltage is in proportion to the rate of change of the input voltage. A high pass RC circuit.

Diffraction. The deviation of an electromagnetic wavefront from the path predicted by geometric optics when the wavefront interacts with, *i.e.*, is restricted by, a physical object such as an opening (aperture) or an edge. *Note:* Diffraction is usually most noticeable for openings of the order of a wavelength. However, diffraction may still be important for apertures many orders of magnitude larger than the wavelength.

Diffraction Grating. An array of fine, parallel, equally spaced grooves ("rulings") on a reflecting or transparent substrate, which grooves result in diffractive and mutual interference effects that concentrate reflected or transmitted electromagnetic energy in discrete directions, called *"orders,"* or *"spectral orders." Note 1:* The groove dimensions and spacings are on the order of the wavelength in question. In the optical regime, in which the use of diffraction gratings is most common, there are many hundreds, or thousands, of grooves per millimeter. *Note 2:* Order zero corresponds to direct transmission or specular reflection. Higher orders result in deviation of the incident beam from the direction predicted by geometric (ray) optics. *Note 3:* Because the angle of deviation of the diffracted beam is wavelength-dependent, a diffraction grating is dispersive, *i.e.*, it separates the incident beam spatially into its constituent wavelength components, producing a spectrum. *Note 4:* The spectral orders produced by diffraction gratings may overlap, depending on the spectral content of the incident beam and the number of grooves per unit distance on the grating. The higher the spectral order, the greater the overlap into the next-lower order. *Note 5:* By controlling the cross-sectional shape of the grooves, it is possible to concentrate most of the diffracted energy in the order of interest. This technique is called *"blazing."*

Diffuse Insolation. Sunlight received indirectly as a result of scattering due to clouds, fog, haze, dust, or other obstructions in the atmosphere. Opposite of direct insolation.

Diffusion. Tendency of conduction band electrons to wander across a pn junction to combine with valence band holes.

Diffusion Furnace. Furnace used to make junctions in semiconductors by diffusing dopant atoms into the surface of the material.

Diffusion Length. The mean distance a free electron or hole moves before recombining with another hole or electron.

Digital. 1. Relating to devices or circuits that have outputs of only two discrete levels. Examples: 0 or 1, high or low, on or off, true or false etc.

2. The newest form of wireless communications that takes all voice transmissions and converts them to computer language (zeros and ones, or "binary" language) and then reconstructs them into the original voice format at the other end. More secure than its original sibling, analog, and also relatively impervious to static or fading signals.

Digital Access Cross-connect System (DACS). A network central office device which enables any T1 facility or any of a T1's 24 channels to be switched or cross-connected to one or more T1 (DS1), DS3 or other high-order facilities.

Digital Facsimile Equipment. Facsimile equipment that digitally encodes the picture signal, *i.e.*, encodes the baseband signal resulting from scanning the object. *Note:* The facsimile equipment output may be either (a) analog, as defined by CCITT Group 3 protocol, or (b) digital, as defined by CCITT Group 4, STANAG 5000 Type I, and STANAG 5000 Type II protocols.

Digital Milliwatt. 1. In digital telephony, a test signal consisting of eight 8-bit words corresponding to one cycle of a sinusoidal signal approximately 1 kHz in frequency and one milliwatt, rms, in power. *Note 1:* The digital milliwatt is stored in ROM. A continuous signal of arbitrary length, *i.e.*, an indefinite number of cycles, may be realized by continually reading out and concatenating the stored information into a data stream to be converted into analog form. *Note 2:* The digital milliwatt is used in lieu of separate test equipment. It has the advantage of being tied in frequency and amplitude to the relatively stable digital clock signal and power (voltage) supply, respectively, that are used by the digital channel bank. 2. A digital signal that is the coded representation of a 0-dBm, 1000-Hertz sine wave.

Digital Signal. Signal represented by a sequence of binary bits representing ones and zeros.

Digital Switch. A switch that performs time-division-multiplexed switching of digitized signals. *Note 1:* When used with analog inputs, analog-to-digital and digital-to-analog conversion are required. These functions may be performed by the digital switch. *Note 2:* Implementation is accomplished by the interchange of time slots between input and output ports on a sequential basis under the direction of control systems. The control systems may be automatic, semiautomatic, or manual.

Digital Transmission System. A transmission system in which (a) all circuits carry digital signals and (b) the signals are combined into one or more serial bit streams that include all framing and supervisory signals. *Note:* A-D/D-A conversion, if required, is accomplished external to the system.

Digitizer. 1. A device that converts an analog signal into a digital representation of the analog signal. *Note:* A digitizer usually samples the analog signal at a constant sampling rate and encodes each sample into a numeric representation of the amplitude value of the sample. 2. A device that converts the position of a point on a surface into digital coordinate data.

Diode. A two terminal device that conducts in only one direction.

DIP. Abbreviation for "dual in line package."

Direct Coupling. Where the output of an amplifier is connected directly to the input of another amplifier or to a load. Also known as DC coupling because DC signals are not blocked.

Direct Current (DC). Electric current in which electrons flow in one direction only. Opposite of alternating current.

Direct Current Signaling (DX signaling). In telephony, a method whereby the signaling circuit E & M leads use the same cable pair(s) as the voice circuit and no filter is required to separate the control signals from the voice transmission.

Direct Gain. In direct-gain buildings, sunlight directly enters the home through the windows and is absorbed and stored in massive floors or walls. These buildings are elongated in the east-west direction, and most of their windows are on the south side. The area devoted to south windows varies throughout the country. It could be as much as 20% of the floor area in sunny cold climates, where advanced glazings or moveable insulation are recommended to prevent heat loss at night. These buildings have high insulation levels and added thermal mass for heat storage.

Direct Insolation. Sunlight falling directly upon a collector. Opposite of diffuse insolation.

Disaggregation. The splitting of a spectrum license into two or more licenses of fewer frequencies.

Discharge. Release of energy stored in either a battery or a capacitor.

Discharge Rate. The rate, usually expressed in amperes or time, at which electrical current is taken from the battery.

DISCO II. FCC proceeding on domestic-international satellite service consolidation.

Disconnect Switch. In a power system, a switch used for closing, opening, or changing the connections in a circuit or system or for purposes of isolation. *Note:* It has no interrupting rating and is intended to be operated only after the circuit has been opened by some other means, such as by a circuit breaker or variable transformer.

Discrete Component. Package containing only a single component as opposed to an integrated circuit containing many components in a single package.

Dispersion. Any phenomenon in which the velocity of propagation of an electromagnetic wave is wavelength dependent. *Note 1:* In communication technology, *"dispersion"* is used to describe any process by which an electromagnetic signal propagating in a physical medium is degraded because the various wave components (*i.e.,* frequencies) of the signal have different propagation velocities within the physical medium. *Note 2:* In an optical fiber, there are several significant dispersion effects, such as material dispersion, profile dispersion, and waveguide dispersion, that degrade the signal. *Note 3:* In optical fiber communications, the incorrect terms *"multimode dispersion"* and *"intermodal dispersion"* should not be used as synonyms for the correct term *Amultimode distortion.*

Dissipation. Release of electrical energy in the form of heat.

Distortion. 1. In a system or device, any departure of the output signal waveform from that which should result from the input signal waveform=s being operated on by the system=s specified, *i.e.,* ideal, transfer function. *Note:* Distortion may result from many mechanisms. Examples include nonlinearities in the transfer function of an active device, such as a vacuum tube, transistor, or operational amplifier. Distortion may also be caused by a passive component such as a coaxial cable or optical fiber, or by inhomogeneities, reflections, *etc.*, in the propagation path. 2. In start-stop teletypewriter signaling, the shifting of the significant instants of the signal pulses from their proper positions relative to the beginning of the start pulse. *Note:* The magnitude of the distortion is expressed in percent of an ideal unit pulse length.

Distributed Capacitance. Any capacitance other than that within a capacitor. For example, the capacitance between adjacent turns of wire in a coil.

Distributed Inductance. Any inductance other than that within an inductor. Example inductance in any conductor.

Distributed Power. Generic term for any power supply located near the point where the power is used. Opposite of central power. See 'stand-alone';'remote site'.

Distributed Systems. Systems that are installed at or near the location where the electricity is used, as opposed to central systems that supply electricity to grids. A residential photovoltaic system is a distributed system.

Disturbance Voltage. An unwanted voltage induced in a system by natural or man-made sources. *Note:* In telecommunications systems, the disturbance voltage creates currents that limit or interfere with the interchange of information. An example of a disturbance voltage is a voltage that produces (a) false signals in a telephone, (b) noise in a radio receiver, or (c) distortion in a received signal.

Diversity (transmission use). In telecommunications, reliability may be improved by use of various forms of *diversity* within a system. Techniques may include time, space, frequency, route, path or other forms of diversity. In addition, redundant equipment, power and other systems can further increase reliability. Different degrees of availability (or reliability) may be achieved depending upon the nature of the diversity used.

Diversity Combiner. A circuit or device for combining two or more signals carrying the same information received via separate paths or channels with the objective of providing a single resultant signal that is superior in quality to any of the contributing signals.

Diversity Reception. Radio reception in which a resultant signal is obtained by combining or selecting signals, from two or more independent sources, that have been modulated with identical information-bearing signals, but which may vary in their fading characteristics at any given instant. *Note 1:* Diversity reception is used to minimize the effects of fading. *Note 2:* The amount of received signal improvement when using diversity reception is directly dependent on the independence of the fading characteristics.

Diversity Transmission. Radio communication using a reception technique in which a resultant signal is obtained by combining signals (a) that originate from two or more independent sources that have been modulated with identical information-bearing signals and (b) that may vary in their transmission characteristics at any given instant. *Note 1:* Diversity transmission and reception are used to obtain reliability and signal improvement by overcoming the effects of fading, outages, and circuit failures. *Note 2:* When using diversity transmission and reception, the amount of received signal improvement depends on the independence of the fading characteristics of the signal as well as circuit outages and failures.

DLCI. Data Link Connection Identifiers (frame-relay use).

DMOS. *Degradation Mean Opinion Score,* an ITU-T P.800 subjective listening test where the test subjects compare two signals, first the reference, then the signal passed through the device under test.

DNS. *Domain Name System,* an automatic Internet phone book, translating an easy-to-remember domain name like heiconsulting.com to its Internet address like 69.6.238.89 and *vice-versa.*

DOD. 'Depth of Discharge,' from 100-percent state of charge (SOC), in a battery or battery system.

Dolby Digital 5.1. Formerly known as AC-3, Dolby Digital® provides efficient encoding and noise reduction for high-quality surround sound™ applications including home entertainment, theatre and broadcasting, including DBS and DVD source material. Dolby Digital requires appreciable bandwidth, 448 Kbps being typical, a reason why it is not available for all material provided by direct broadcast satellite.

Domain. A moveable magnetized area in a magnetized material. Also known as magnetic domain.

Donor. An n-type dopant that puts an additional electron into an energy level very near the conduction band; this electron is easily exited into the conduction band where it increases the electrical conductivity over than of an undoped semiconductor.

Donor Atoms. Pentavalent atoms that give up electrons to the conduction band in an N type semiconductor material.

Donor Level. The level that donates conduction electrons to the system.

Dopant. A chemical element (impurity) added in small amounts to an otherwise pure semiconductor material to modify the electrical properties of the material. An n-dopant introduces more electrons. A p-dopant creates electron vacancies (holes).

Doping. The process of adding impurity atoms to intrinsic (pure) silicon or germanium to improve the conductivity of the semiconductor material.

Dot Convention. Standard used with transformer symbols to indicate whether the secondary voltage is in phase or out of phase with the primary voltage.

Double-sideband Reduced Carrier (DSB-RC) Transmission. Transmission in which (a) the frequencies produced by amplitude modulation are symmetrically spaced above and below the carrier and (b) the carrier level is reduced for transmission at a fixed level below that which is provided to the modulator. *Note:* In DSB-RC transmission, the carrier is usually transmitted at a level suitable for use as a reference by the receiver, except for the case in which it is reduced to the minimum practical level, *i.e.*, the carrier is suppressed.

Downlink. The portion of a telecommunications path from a satellite to the ground. Also referred to as the reverse link.

DPCM. Differential Pulse Code Modulation. A digital coding and compression process in which the analog signal is first sampled, and the difference between each sample of the signal and its prior estimated value is quantized and digitally encoded.

Drift. A problem that can develop in tuned amplifiers when the frequency of the tuned circuit changes due to temperature or component aging.

Dropping Resistor. Resistor whose value has been chosen to drop or develop a given voltage.

Drive Test. A method of taking signal strength measurements in a cellular coverage area.

Driver Distraction. A key concern of transportation and law enforcement agencies, wireless carriers and other entities, driver distraction is a diversion of attention caused by many factors. Factors can include changing a compact disc or cassette tape, tuning a broadcast radio dial, eating or drinking or talking on cellular phones. The wireless industry has launched a consumer education campaign to emphasize safe cellular use while driving.

Dropped Call. A wireless call that is unintentionally disconnected due to a system problem, lack of channel availability or dead spot in coverage.

Dry Cell. DC voltage generating chemical cell using a non liquid (paste) electrolyte.

DS0. The AT&T Bell System level 0 digital standard for systems operating at 64-Kbps.

DS1. The AT&T Bell System level 1 digital standard for systems operating at 1.544 Mbps and consisting of 24 DS-0 channels.

DS3. The AT&T Bell System level 3 digital standard for systems operating at 44.736 (45) Mbps.

DSL. *Digital Subscriber Line*, a generic term encompassing a family of moderate speed access using a subscriber's existing copper lines.

DSLAM. *Digital Subscriber Line Access Multiplexer*, a central office device which terminates all the customer xDSL lines.

DSP (Digital Signal Processor). A specialized microprocessor that performs mathematical operations on a data stream in real time to produce a second (modified) data stream.

DSU. *Digital Service Unit*, see CSU.

DSX-1. *Digital Signal Cross-connect level 1*, a DS0 or T1 switch and interface specification.

DTE. A configuration protocol that identifies Data Terminal Equipment; the type of data communications connection at equipment connecting to a DTE device, compare DCE.

DTH. *Direct-To Home*. Simplex satellite broadcast services like DirecTV, Echostar and ExpressVu, transmitting directly to end-users. Similarly, MMDS or wireless cable can provide service from a fixed terrestrial transmitter site often including a return data or audio path. No wired system is needed, making these technologies ideal for sparsely populated areas.

DTMF. Dual-Tone Multi-Frequency, the method of telephony signaling used by all modern telephones, modems and similar equipment.

DTV (Digital Television). The next generation of video and audio technology for TV broadcasters. Often synonymous with high-definition television or HDTV. According to the 1996 Telecommunications Act, broadcasters can offer 'ancillary and supplementary services' such as paging and data in addition to video programming.

Dual Access. 1. The connection of a user to two switching centers by separate access lines using a single message routing indicator or telephone number. 2. In satellite communications, the transmission of two carriers simultaneously through a single communication satellite repeater.

Dual Band. Describes a handset that works on 800 MHz cellular and 1900 MHz PCS frequencies.

Dual Communications Interface. An interface that allows simultaneous transmission of two halves of a set of communications data over two similar communication interfaces. A familiar example is ISDN-BRI (2B+D), a connection using two switched 56- or 64- Kbps transmission circuits. See also H.244.

Dual In-line Package (DIP). An electronic package with a rectangular housing and a row of pins along each of two opposite sides. *Note:* DIP packages may be used for integrated circuits, or for discrete components, such as resistors or toggle switches. An example of a DIP is a microcircuit package with two rows of seven vertical leads that is specially designed for mounting on a printed circuit board.

Dual Mode. Describes a handset that works on both analog and digital networks.

Dual Tone Multi Frequency (DTMF). The sounds made by a phone's keypad when a button is pressed. Each button emits a sound that is actually the combination of two specific sounds in order to minimize the possibility of an incorrect signal being received by the equipment listening to the press of the buttons.

Dual Trace Oscilloscope. Oscilloscope that can simultaneously display two signals.

Duct. 1. In interfacility cabling, a conduit, which may be direct-earth buried or encased in concrete, used to enclose communications or power cables. *Note:* For maximum resistance to rodent attack, direct-earth-buried conduit should have an outside diameter equal to or greater than 6 cm (2.25 in.).

Dynamic. Relating to conditions that are changing or in motion.

E

E1. Abbreviation for the European equivalent of DS1 (T1) (ITU-T Rec. G.703 and G.732), 2.048 Mbps.

E3. Abbreviation for the European equivalent of DS3, 34.368 Mbps.

Earth Electrode Subsystem. A network of electrically interconnected rods, plates, mats, or grids, installed and connected, for the purpose of establishing a low-resistance contact with earth.

E-bend. A smooth change in the direction of the axis of a waveguide, throughout which the axis remains in a plane parallel to the direction of electric E-field (transverse) polarization. *Synonym* E-plane bend.

EBU. European Broadcasting Union. Headquartered in Geneva CH and with its technical center in Brussels BE, the EBU is an association of European and world broadcasters, including the Eurovision Network. The EBU provides regulatory, legal and technical services, as well as programming to its members.

Echo. 1. A wave that has been reflected by a discontinuity in the propagation medium. 2. A wave that has been reflected or otherwise returned with sufficient magnitude and delay to be perceived. *Note 1:* Echoes are frequently measured in dB relative to the directly transmitted wave. *Note 2:* Echoes may be desirable (as in radar) or undesirable (as in telephone systems). 3. In computing, to print or display characters (a) as they are entered from an input device, (b) as instructions are executed, or (c) as retransmitted characters received from a remote terminal. 4. In computer graphics, the immediate notification of the current values provided by an input device to the operator at the display console.

Echo Attenuation. In a communication circuit (4- or 2-wire) in which the two directions of transmission can be separated from each other, the attenuation of echo signals that return to the input of the circuit under consideration. *Note:* Echo attenuation is expressed as the ratio of the transmitted power to the received echo power in dB.

Echo Cancellation. In a system, the reduction of the power level of an echo or the elimination of an echo. *Note:* Echo cancellation is usually an active process in which echo signals are measured and canceled or eliminated by combining an inverted signal with the echo signal.

Echo Return Loss Enhancement (ERLE). A figure of merit for acoustic and telephony echo cancellers, expressed in dB. An echo canceller electronically enhances the apparent loss between tends of a telecommunications circuit (TEC) or between the room loudspeaker and the microphones (AEC).

E-core. Laminated form in the shape of the letter "E", onto which inductors and transformers are wound.

ECO Test (Effective Competitive Opportunities Test). Developed to determine whether U.S. carriers enjoy effective competition in a foreign carrier's market before granting U.S. market access to that foreign carrier. Following the Basic Telecom Agreement signed by World Trade Organization members in February, 1997, the FCC decided not to apply the ECO test to WTO members.

EDACS (Enhanced Digital Access Communications System). A private radio or specialized mobile radio network designed by Ericsson Inc.

Eddy Currents. Currents induced into a conducting core due to the changing magnetic field. Eddy currents produce heat which is a loss of power and lowers the efficiency of an inductor.

EDGE (Enhanced Data Rates for Global (or GSM) Evolution). An advanced technology for GSM and TDMA networks that may offer wireless data access speeds of up to 384 kilobits per second in end-user devices.

Edge-emitting LED. An LED that has a physical structure superficially resembling that of an injection laser diode, operated below the lasing threshold and emitting incoherent light. *Note:* Edge-emitting LEDs have a relatively small beam divergence, and thus are capable of launching more optical power into a given fiber than are the conventional surface-emitting LEDs.

Edge-defined Film-fed Growth (EFG). A method for making sheets of polycrystalline silicon in which molten silicon is drawn upward by capillary action through a mold.

EEPROM. Electrically-Erasable Programmable Read-Only Memory. A non-volatile but electrically alterable form of semiconductor storage. These devices are present in some telecommunications equipment such as codecs, permitting long-term storage of configuration, dialing and related information for easy recall by the user. A fast, reliable EEPROM version is popularly known as FLASH memory.

Efficiency. The amount of power delivered to the load of an amplifier as a percentage of the power required from the power supply.

Echo Suppression. The insertion of mild attenuation in audio transmit and/or receive signal path to eliminate howling caused by excessive closed-loop gain. Echo suppression may also be used with TECs and AECs to attenuate any residual echo signals.

Effective Input Noise Temperature. The source noise temperature in a two-port network or amplifier that will result in the same output noise power, when connected to a noise-free network or amplifier, as that of the actual network or amplifier connected to a noise-free source. *Note:* If *F* is the noise figure numeric and 290 K the standard noise temperature, then the effective noise temperature is given by $T_n = 290(FB1)$.

Effective Mode Volume. For an optical fiber, the square of the product of the diameter of the near-field pattern and the sine of the radiation angle of the far-field pattern. The diameter of the near-field radiation pattern is defined here as the full width at half maximum and the radiation angle at half maximum intensity. *Note:* Effective mode volume is proportional to the breadth of the relative distribution of power amongst the modes in a multimode fiber. It is not truly a spatial volume but rather an *"optical volume"* equal to the product of area and solid angle.

Effective Monopole Radiated Power (e.m.r.p.). The product of the power supplied to the antenna and its gain relative to a short vertical antenna in a given direction.

Effective Power. In alternating-current power transmission and distribution, the product of the rms voltage and amperage, *i.e.*, the apparent power, multiplied by the power factor, *i.e.*, the cosine of the phase angle between the voltage and the current. *Note:* Only effective power, *i.e.*, the actual power delivered to or consumed by the load, is expressed in watts. Apparent power is properly expressed only in volt-amperes, never watts. *Synonym* true power.

EIA. Electronic Industries Association. A North American standards-setting body.

EIA Interface. Any of a number of equipment interfaces compliant with voluntary industry standards developed by the Electronic Industries Association (EIA) to define interface parameters. *Note 1:* Some of the EIA interface standards have been adopted by the Federal Government as Federal Standards or Federal Information Processing Standards. *Note 2:* The telecommunication-standards-developing bodies of the EIA are now part of the Telecommunications Industry

Association (TIA), and the standards are designated TIA/EIA-XXX.

Eighth Floor. The eighth floor of the FCC headquarters at 1919 M St., Washington, D.C., where the commissioners' offices are located. Commonly used to refer collectively to the commissioners. May become obsolete when the commission relocates to The Portals.

802.11. The Institute of Electrical and Electronics Engineers standard for wireless local area network interoperability.

Electrical Length. 1. Of a transmission medium, its length expressed as a multiple or submultiple of the wavelength of a periodic electromagnetic or electrical signal propagating within the medium. *Note 1:* The wavelength may be expressed in radians or in artificial units of angular measure, such as degrees. *Note* 2: In both coaxial cables and optical fibers, the velocity of propagation is approximately two-thirds that of free space. Consequently, the wavelength will be approximately two-thirds that in free space, and the electrical length, approximately 1.5 times the physical length. 2. Of a transmission medium, its physical length multiplied by the ratio of (a) the propagation time of an electrical or electromagnetic signal through the medium to (b) the propagation time of an electromagnetic wave in free space over a distance equal to the physical length of the medium in question. *Note:* The electrical length of a physical medium will always be greater than its physical length. For example, in coaxial cables, distributed resistances, capacitances and inductances impede the propagation of the signal. In an optical fiber, interaction of the lightwave with the materials of which the fiber is made, and fiber geometry, affect the velocity of propagation of the signal. 3. Of an antenna, the effective length of an element, usually expressed in wavelengths. *Note 1:* The electrical length is in general different from the physical length. *Note 2:* By the addition of an appropriate reactive element (capacitive or inductive), the electrical length may be made significantly shorter or longer than the physical length.

Electrically Powered Telephone. A telephone in which the operating power is obtained either from a battery located at the telephone, *i.e.*, a local battery, or from a telephone central office, *i.e.*, a common battery.

Electric Charge. Electric energy stored on the surface of a material. Also known as a static charge.

Electric Circuit. Path followed by electrons from a power source (generator or battery) through an external line (including devices that use the electricity) and returning through another line to the source.

Electric Current. A flow of electrons; electricity.

Electric Field. The effect produced by the existence of an electric charge, such as an electron, ion, or proton, in the volume of space or medium that surrounds it. *Note:* Each of a distribution of charges contributes to the whole field at a point on the basis of superposition. A charge placed in the volume of space or in the surrounding medium has a force exerted on it.

Electrical Grid. An integrated system of electricity distribution, usually covering a large area.

Electricity. Science states that certain particles possess a force field or charge. The charge possessed by an electron is negative while the charge possessed by a proton is positive. Electricity can be divided into two groups, static and dynamic. Static electricity deals with charges at rest and dynamic electricity deals with charges in motion.

Electric Polarization. A displacement of bound charges in a dielectric when placed in an electric field.

Electroacoustic Transducer. Device that produces an energy transfer from electric to acoustic (sound) or from acoustic to electric. Examples include a microphone, earphones and loudspeakers.

Electrochemical Recording. Facsimile recording by means of a chemical reaction brought about by the passage of a signal-controlled current through the sensitized portion of the record sheet.

Electrodeposition. Electrolytic process in which a metal is deposited at the cathode from a solution of its ions.

Electroluminescence. Nonthermal conversion of electrical energy into light. *Note 1:* Electroluminescence is distinguished from incandescence, which is a thermal process. *Note 2:* One example of electroluminescence is the photon emission resulting from electron-hole recombination in a pn junction, as in a light-emitting diode (LED).

Electrolyte. A liquid conductor of electricity. In batteries, usualy H_2SO_4, sulfuric acid, but may be any number of things. Seawater is the mostcommon electrolyte in the world - and by suspending a zinc and a steel sheet in it, you can get a little electricity.

Electrolytic Capacitor. A capacitor having an electrolyte between the two plates. A thin layer of oxide is deposited on only the positive plate. The oxide acts as the dielectric for the capacitor. Electrolytic capacitors are polarized and so must be connected in correct polarity to prevent breakdown.

Electrolytic Recording. Electrochemical facsimile recording in which the recorded copy is made by the passage of a signal-controlled current through an electrolyte which causes metallic ions to be deposited, thus forming an image of the object.

Electromagnet. A coil of wire usually wound on a soft iron or steel core. When current is passed through the coil a magnetic field is generated. The core provides an easy path for the magnetic lines of force. This concentrates the field in the core.

Electromagnetic Communication. Use of an electromagnetic wave to pass information between two points. Also called wireless communication.

Electromagnetic Compatibility (EMC). 1. Electromagnetic compatibility is the condition which prevails when telecommunications equipment is performing its individually designed function in a common electromagnetic environment without causing or suffering unacceptable degradation due to unintentional electromagnetic interference to or from other equipment in the same environment. 2. The ability of systems, equipment, and devices that utilize the electromagnetic spectrum to operate in their intended operational environments without suffering unacceptable degradation or causing unintentional degradation because of electromagnetic radiation or response. It involves the application of sound electromagnetic spectrum management; system, equipment, and device design configuration that ensures interference-free operation; and clear concepts and doctrines that maximize operational effectiveness.

Electromagnetic Emission Control. The control of friendly electromagnetic emissions, such as radio and radar transmissions, for the purpose of preventing or minimizing their use by unintended recipients.

Electromagnetic Environment (EME). 1. For a telecommunications system, the spatial distribution of electromagnetic fields surrounding a given site. *Note:* The electromagnetic environment may be expressed in terms of the spatial and temporal distribution of electric field strength (volts/meter), irradiance (watts/meter^2), or energy density (joules/meter^3). 2. The resulting product of the power and time distribution, in various frequency ranges, of the radiated or conducted electromagnetic emission levels that may be encountered by a military force, system, or platform when performing its assigned mission in its intended operational environment. It is the sum of electromagnetic interference; electromagnetic pulse; hazards of electromagnetic radiation to personnel, ordnance, and volatile materials; and natural phenomena effects of lightning and p-static.

Electromagnetic Induction. Voltage produced in a coil due to relative motion between the coil and magnetic lines of force.

Electromagnetic Interference (EMI). Any electromagnetic disturbance that interrupts, obstructs, or otherwise degrades or limits the effective performance of electronics/electrical equipment. It can be induced intentionally, as in some forms of electronic warfare, or unintentionally, as a result of spurious emissions and responses, intermodulation products, and the like. *Synonym* radio frequency interference.

Electromagnetic Interference (EMI) Control. The control of radiated and conducted energy such that emissions that are unnecessary for system, subsystem, or equipment operation are reduced, minimized, or eliminated. *Note:* Electromagnetic radiated and conducted emissions are controlled regardless of their origin within the system, subsystem, or equipment. Successful EMI control with effective susceptibility control leads to electromagnetic compatibility.

Electromagnetic Intrusion. The intentional insertion of electromagnetic energy into transmission paths in any manner, with the objective of deceiving operators or of causing confusion.

Electromagnetic Pulse (EMP). 1. The electromagnetic radiation from a nuclear explosion caused by Compton-recoil electrons and photoelectrons from photons scattered in the materials of the nuclear device or in a surrounding medium. The resulting electric and magnetic fields may couple with electrical/electronic systems to produce damaging current and voltage surges. May also be caused by nonnuclear means. 2. A broadband, high-intensity, short-duration burst of electromagnetic energy. *Note:* In the case of a nuclear detonation, the electromagnetic pulse consists of a continuous frequency spectrum. Most of the energy is distributed throughout the lower frequencies between 3 Hz and 30 kHz.

Electromagnetic Radiation (EMR). Radiation made up of oscillating electric and magnetic fields and propagated with the speed of light. Includes gamma radiation, X-rays, ultraviolet, visible, and infrared radiation, and radar and radio waves.

Electromagnetic Radiation Hazards (RADHAZ *or* EMR hazards). Hazards caused by a transmitter/antenna installation that generates electromagnetic radiation in the vicinity of ordnance, personnel, or fueling operations in excess of established safe levels or increases the existing levels to a hazardous level; or a personnel, fueling, or ordnance installation located in an area that is illuminated by electromagnetic radiation at a level that is hazardous to the planned operations or occupancy. These hazards will exist when an electromagnetic field of sufficient intensity is generated to: (a) induce or otherwise couple currents and/or voltages of magnitudes large enough to initiate electroexplosive devices or other sensitive explosive components of weapon systems, ordnance, or explosive devices; (b) cause harmful or injurious effects to humans and wildlife; (c) create sparks having sufficient magnitude to ignite flammable mixtures of materials that must be handled in the affected area.

Electromagnetic Spectrum. The range of frequencies of electromagnetic radiation from zero to infinity. *Note:* The electromagnetic spectrum was, by custom and practice, formerly divided into 26 alphabetically designated bands. This usage still prevails to some degree. However, the ITU formally recognizes 12 bands, from 30 Hz to 3000 GHz. New bands, from 3 THz to 3000 THz, are under active consideration for recognition.

Electromagnetic Survivability. The ability of a system, subsystem, or equipment to resume functioning without evidence of degradation following temporary exposure to an adverse electromagnetic environment. *Note:* The system, subsystem, or equipment performance may be degraded during exposure to the adverse electromagnetic environment, but the system will not experience permanent damage, such as component burnout, that will prevent proper operation when the adverse electromagnetic environment is removed.

Electromagnetic Vulnerability (EMV). The characteristics of a system that cause it to suffer a definite degradation (incapability to perform the designated mission) as a result of having been subjected to a certain level of electromagnetic environmental effects.

Electromagnetic Wave. 1. Wave that consists of both electric and magnetic variation. 2. A wave produced by the interaction of time-varying electric and magnetic fields.

Electromagnetism. Relates to the magnetic field generated around a conductor when current is passed through it.

Electromechanical Recording. Recording by means of a signal-actuated mechanical device.

Electromechanical Transducer. Device that transforms electrical energy into mechanical energy (electric motor) or mechanical energy into electrical energy (generator).

Electromotive Force (emf). Force that causes the motion of electrons due to potential difference between two points. (voltage)

Electron. Smallest sub atomic particle of negative charge that orbits the nucleus of an atom.

Electron Flow. Electrical current produced by the movement of free electrons towards a positive terminal.

Electronic Assistant. Generic term for telecommunications access systems based on voice-activation or voice-response technology, enabling the user to dial phone numbers or send and receive information such as voice messages or other content. Systems may be network based, such as Wildfire Communications Inc.'s Wildfire, or offered via a service bureau, such as General Magic Inc.'s Portico or Motorola Inc.'s planned Myosphere.

Electronic Deception. 1. The deliberate radiation, reradiation, alteration, suppression, absorption, denial, enhancement, or reflection of electromagnetic energy in a manner intended to convey misleading information and to deny valid information to an enemy or to enemy electronics-dependent weapons. *Note:* Among the types of electronic deception are: (a) manipulative electronic deceptionCActions to eliminate revealing or convey misleading, telltale indicators that may be used by hostile forces; (b) simulative electronic deceptionCActions to represent friendly notional or actual capabilities to mislead hostile forces; (c) imitative electronic deceptionCThe introduction of electromagnetic energy into enemy systems that imitates enemy emissions. 2. Deliberate activity designed to mislead an enemy in the interpretation or use of information received by his electronic systems.

Electronic Switching System (ESS). 1. A telephone switching system based on the principles of time-division multiplexing of digitized analog signals. *Note:* An electronic switching system digitizes analog signals from subscribers= loops, and interconnects them by assigning the digitized signals to the appropriate time slots. It may also interconnect digital data or voice circuits. 2. A switching system with major devices constructed of semiconductor components. *Note:* A semi-electronic switching system that has reed relays or crossbar matrices, as well as semiconductor components, is also considered to be an ESS.

Electron Volt. An energy unit equal to the energy an electron acquires when it passes through a potential difference of one volt; it is equal to 1.602 x 10-19 volt.

Electro-optic Effect. Any one of a number of phenomena that occur when an electromagnetic wave in the optical spectrum interacts with an electric field, or with matter under the influence of an electric field. *Note 1:* Two of the most important electro-optic effects having application as modulation mechanisms in optical communication are the Kerr effect and the Pockels effect, in which birefringence is induced or modified in a liquid (Kerr effect) or solid (Pockels effect). *Note 2:* The term *"electro-optic"* is often erroneously used as a synonym for *"optoelectronic."*

Electro-optic Modulator. An optical device in which a signal-controlled element is used to modulate a beam of light. *Note 1:* The modulation may be imposed on the phase, frequency, amplitude, or direction of the modulated beam. *Note 2:* Modulation bandwidths into the gigahertz range are possible using laser-controlled modulators.

Electro-optics. The technology associated with those components, devices and systems which are designed to interact between the electromagnetic (optical) and the electric (electronic) state. *Note 1:* The operation of electro-optic devices depends on modification of the refractive index of a material by electric fields. *Note 2:* In a Kerr cell, the refractive index change is proportional to the square of the electric field, and the material is usually a liquid. *Note 3:* In a Pockels cell, the refractive index change varies linearly with the electric field, and the material is a crystal. *Note 4: "Electro-optic"* is often erroneously used as a synonym for *"optoelectronic"*.

Electrostatic. Related to static electric charge.

Electrostatic Field. Force field produced by static electrical charges.

Elliptical Polarization. In electromagnetic wave propagation, polarization such that the tip of the electric field vector describes an ellipse in any fixed plane intersecting, and normal to, the direction of propagation. *Note 1:* An elliptically polarized wave may be resolved into two linearly polarized waves in phase quadrature with their polarization planes at right angles to each other. *Note 2:* Circular and linear polarization are special cases of elliptical polarization.

Emission. 1. Electromagnetic energy propagated from a source by radiation or conduction. *Note:* The emission may be either desired or undesired and may occur anywhere in the electromagnetic spectrum. 2. Radiation produced, or the production of radiation, by a radio transmitting station. For example, the energy radiated by the local oscillator of a radio receiver would not be an emission but a radiation.

Emitter. The semiconductor region from which charge carriers are injected into the base of a bipolar junction transistor.

Emitter Feedback. Coupling from the emitter output to the base input of a bipolar junction transistor.

Emitter Follower. A common collector amplifier. Has a high current gain, high input impedance and low output impedance.

Emphasis. In FM transmission, the intentional alteration of the amplitude-vs.-frequency characteristics of the signal to reduce adverse effects of noise in a communication system. *Note:* The high-frequency signal components are emphasized to produce a more equal modulation index for the transmitted frequency spectrum, and therefore a better signal-to-noise ratio for the entire frequency range.

Encryption. 1. The process of "scrambling" a message such as a digital phone signal to prevent it from being read by unauthorized parties.

2. One of a number of approaches to convert "plain text" or data into information that is unreadable except by the intended recipients. Adjectives like "weak" and "strong" are often used to further describe the encryption systems. Similarly, under government or military applications, encryption may be described as providing "privacy" or "secrecy".

Energized. Being electrically connected to a voltage source so the device is activated.

Energy. The ability to do work. Stored energy becomes working energy when we use it.

Energy Audit. A survey that shows how much energy you use in your house or apartment. It will help you find ways to use less energy.

Energy Contribution Potential. Recombination occurring in the emitter region of a photovoltaic cell.

Energy Density. The ratio of energy available from a battery to its volume (Wh/1) or mass (Wh/kg).

Energy Levels. The energy represented by an electron in the band model of a substance.

Energy Payback Time. The time required for any energy producing system or device to produce as much energy as was required in its manufacture. For solar electric panels, this is about 16-20 months.

Engineering Notation. A floating point system in which numbers are expressed as products consisting of a number greater than one multiplied by an appropriate power of ten that is some multiple of three.

Enhancement-mode MOSFET. A field effect transistor in which there are no charge carriers in the channel when the gate source voltage is zero.

E911 (Enhanced 911). 911 service becomes E911 when automatic number identification and automatic location information is provided to the 911 operator.

***E/N* Ratio.** In the transmission of a pulse of an electromagnetic wave representing a bit, the ratio of (a) the energy in each bit, E, to (b) the noise energy density per hertz, N. *Note:* E is usually expressed in joules per bit and N is usually expressed in watts per hertz. Thus, the E/N ratio is hertz-seconds per bit. A joule is a watt-second and a hertz is a cycle per second. Thus, the E/N ratio is actually cycles per bit. However, if a cycle is a bit, then the E/N ratio is dimensionless.

Environment. All the natural and living things around us. The earth, air, weather, plants, and animals all make up our environment.

Epitaxial Growth. The growth of one crystal on the surface of another crystal. The growth of the deposited crystal is oriented by the lattice structure of the original crystal.

EPROM. Electrically Programmable Read-Only Memory. A non-volatile, electrically-write-once semiconductor storage device, useful for storage of parameters which normally do not need updating during the life of the product or system.

Equal Access. Allows long-distance carriers the opportunity to bid for access to local carrier's customer's LD subscription.

Equivalent Noise Temperature. The temperature, usually expressed in kelvins, of a hypothetical matched resistance at the input of an assumed noiseless device, such as a noiseless amplifier, that would account for the measured output noise.

Equivalent Resistance. Total resistance of all the individual resistances in a circuit.

ERMES (European Radio Messaging System). A paging system used in Europe and other parts of the world.

ESF. Extended Superframe Format. A network control and framing method used by common carriers that incorporates control and information data in the transmitted bit stream. When connecting networks and equipment, it is important that both the equipment and the network support the same appropriate protocols.

ESMR (Enhanced Specialized Mobile Radio). Digital SMR networks, usually referring to Nextel Communications Inc., which provide dispatch, voice, messaging and data services.

ESN (Electronic Serial Number). The unique identification number embedded in a wireless phone by the manufacturer. Each time a call is placed, the ESN is automatically transmitted to the base station so the wireless carrier's mobile switching office can check the call's validity. The ESN cannot be altered in the field. The ESN differs from the mobile identification number, which is the wireless carrier's identifier for a phone in the network. MINs and ESNs can be electronically checked to help prevent fraud.

ETSI (European Telecommunications Standards Institute). A standards-setting body in Europe.

EVA (Ethylene Vinyl Acetate). An encapsulant used between the glass cover and the solar cells in PV modules. It is durable, transparent, resistant to corrosion, and flame retardant.

Evanescent Field. In a waveguide, a time-varying field having an amplitude that decreases monotonically as a function of transverse radial distance from the waveguide, but without an accompanying phase shift. *Note 1:* The evanescent field is coupled, *i.e.,* bound, to an electromagnetic wave or mode propagating inside the waveguide. *Note 2:* The evanescent field is a surface wave. *Note 3:* In fiber optics, the evanescent field may be used to provide coupling to another fiber.

Ex parte. Statements, meetings or filings that are made outside of an official comment-and-replay period. They must be reported and a summary of them made available in the public record.

Extrinsic Semiconductor. The product of doping a pure semiconductor.

F

FAC. Certified frequency advisory committee, also known as a frequency coordinator.

Facsimile Converter. 1. In a facsimile receiver, a device that changes the signal modulation from frequency-shift keying (FSK) to amplitude modulation (AM). 2. In a facsimile transmitter, a device that changes the signal modulation from amplitude modulation (AM) to frequency-shift keying (FSK).

Facsimile Signal Level. In a facsimile system, the signal level at any point in the system. *Note 1:* The facsimile signal level is used to establish the operating levels. *Note 2:* The facsimile signal level is usually expressed in dB with respect to some standard value, such as 1 mW (milliwatt), *i.e.*, 0 dBm.

Fall Time. Time it takes the falling edge of a pulse to go from 90% of peak voltage to 10% of peak voltage.

Farad. The basic unit of capacitance.

Faraday Effect. A magneto-optic effect in which the polarization plane of an electromagnetic wave is rotated under the influence of a magnetic field parallel to the direction of propagation. *Note:* The Faraday effect may be used to modulate a lightwave.

Fault Protection Subsystem. In a facility power distribution system, the subsystem that provides a direct path from each power sink to the earth electrode subsystem. *Note:* The fault protection subsystem is usually referred to as a *"green wire."*

Fault-tolerant. A method of making a network system or a computer resistant to software errors, hardware problems or power failures.

FCC (Federal Communications Commission). A US body regulating, approving and licensing radiated electromagnetic signals including broadcasting and telecommunications.

FDDI. *Fiber Distributed Data Interface.* A North American 100-Mbps, reliable dual-ring voice and data building distribution system.

FDM. *Frequency Division Multiplex.*

FDMA. *Frequency-Division, Multiple Access.*

Feature Group D. Local exchange carrier network service that, among other things, lets public-safety dispatch offices receive a 10-digit data stream, including the full call-back number, alongside wireless 911 calls. Offered as a way by which wireless carriers can meet FCC enhanced 911 rules and dispatch offices can overcome their current bandwidth limits.

FEC. *Forward Error Correction.*

Feed. 1. To supply a signal to the input of a system, subsystem, equipment, or component, such as a transmission line or antenna. 2. A coupling device between an antenna and its transmission line. *Note:* A feed may consist of a distribution network or a primary radiator. 3.. A transmission facility between (a) the point of origin of a signal, such as is generated in a radio or television studio, and (b) the head-end of a distribution facility, such as a broadcasting station in a network. 4. Pertaining to the function of inserting one thing into another, such as in a feed horn, paper feed, card feed, and line feed.

Feedback. A portion of the output signal of an amplifier which is connected back to the input of the same amplifier.

Feedback Amplifier. An amplifier with an external signal path from its output back to its input.

Fermi Level. Energy level at which the probability of finding an electron is one-half. In a metal, the Fermi level is very near the top of the filled levels in the partially filled valance band. In a semiconductor, the Fermi level is in the band gap.

Ferrite. A powdered, compressed and sintered magnetic material having high resistivity. The high resistance makes eddy current losses low at high frequencies.

Ferrite Bead. Ferrite composition in the form of a bead. Running a wire through the bead increases the inductance of the wire.

Ferrite-core Inductor. An inductor wound on a ferrite core.

Ferrites. Compound composed of iron oxide, a metallic oxide and ceramic. The metal oxides include zinc, nickel, cobalt or iron.

Ferrous. Composed of and or containing iron. A ferrous metal exhibits magnetic characteristics as opposed to non-ferrous material.

FHMA (Frequency Hopping Multiple Access). A digital technology used in Geotek Communications Inc.'s specialized mobile radio network.

Fiber Amplifier. A device that amplifies an optical signal directly, without the need to convert it to an electrical signal, amplify it electrically, and reconvert it to an optical signal. *Note 1:* One type of fiber amplifier uses a doped fiber (*e.g.*, a fiber doped with erbium), which bears the communication signal, and which is optically pumped with a laser having a high-powered continuous output at an optical frequency slightly higher than that of the communication signal. The signal is intensified by Raman amplification. *Note 2:* Because neither optical-electrical conversion nor electrical amplification takes place, this type of amplifier is well suited for a wide variety of applications, both digital and analog. *Note 3:* Because this type of amplifier does not require extraordinary frequency (wavelength) control of the pumping laser, it is relatively simple. *Synonym* Raman amplifier.

Fiber Optics (FO). The branch of optical technology concerned with the transmission of light through fibers made of transparent materials such as glasses and plastics. [2196] *Note 1:* Telecommunications applications of fiber optics use flexible low-loss fibers, using a single fiber per optical path. Present-day plastic fibers have losses that are too high for telecommunications applications. *Note 2:* Various industrial and medical applications of fiber optics, such as endoscopes, use flexible fiber bundles in which individual fibers are spatially aligned, permitting optical relay of an image. *Note 3:* Some specialized industrial applications use rigid (fused) aligned fiber bundles for image transfer; such as in the fiber optics faceplates used on some cathode ray rubes (CRTs) to "flatten" the image.

Field Effect Transistor (FET). A voltage controlled transistor in which the source to drain conduction is controlled by gate to source voltage.

Field Intensity. The irradiance of an electromagnetic wave under specified conditions. *Note:* Field intensity is usually expressed in watts per square meter.

Field Strength. The magnitude of an electric, magnetic, or electromagnetic field at a given point. *Note:* The field strength of an electromagnetic wave is usually expressed as the rms value of the electric field, in volts per meter. The field strength of a magnetic field is usually expressed in ampere-turns per meter or in oersteds. *Synonym* radio field intensity.

Field Subsampling. See *Spatial Filtering.*

Field Wire. A flexible insulated wire used in field telephone and telegraph systems. *Note 1:* WD-1 and WF-16 are types of field wire. *Note 2:* Field wire usually contains conductors and high-tensile-strength strands serving as strength members.

Filament. Thin thread of carbon or tungsten which produces heat or light with the passage of current.

Filled Cable. A cable that has a nonhygroscopic material, usually a gel, inside the jacket or sheath. *Note 1:* The nonhygroscopic material fills the spaces between the interior parts of the cable, preventing moisture from entering minor leaks in the sheath and migrating inside the cable. *Note 2:* A metallic cable, such as a coaxial cable or a metal waveguide, filled with a dielectric material, is not considered as a filled cable.

Filter. In electronics, a device that transmits only part of the incident energy and may thereby change the spectral distribution of energy: (a) high-pass filters transmit energy above a certain frequency; (b) low-pass filters transmit energy below a certain frequency; (c) bandpass filters transmit energy of a certain bandwidth; (d) band-stop filters transmit energy outside a specific frequency band.

Fill Factor. The ratio of a photovoltaic cell's actual power to its power if both current and voltage were at their maxima. A key characteristic in evaluating cell performance.

Firmware. Data and/or program software for an electronic device stored in a non-volatile form using a semiconductor memory circuit.

5ESS. A registered trademark name for Lucent Technologies Inc.'s electronic switching system.

Flash Memory. A fast, reliable variant of EEPROM or Electrically-Erasable Programmable Read-Only Memory, a non-volatile, electrically alterable form of semiconductor storage present in some telecommunications equipment for long-term storage of configuration, dialing and related information.

Flash ROM. A class of Electrically Alterable Programmable Read Only Memory (EAPROM), the contents of which can be altered in the field, normally without disassembly of the product containing the memory. Maintains data in absence of power.

Flat-plate Photovoltaic Module. An arrangement of photovoltaic cells mounted on a rigid flat surface with the cells exposed freely to incoming sunlight.

Flat-plate PV. Refers to a PV array or module that consists of nonconcentrating elements. Flat-plate arrays and modules use direct and diffuse sunlight, but if the array is fixed in position, some portion of the direct sunlight is lost because of oblique sun-angles in relation to the array.

Flat Weighting. In a noise-measuring set, a noise weighting based on an amplitude-frequency characteristic that is flat over a frequency range that must be stated. *Note 1:* Flat noise power is expressed in dBrn (f_1 - f_2) or in dBm (f_1 - f_2). *Note 2: "3-kHz flat weighting"* and *"15-kHz flat weighting"* are based on amplitude-frequency characteristics that are flat between 30 Hz and the frequency indicated.

FLEX. A Motorola Inc.-licensed protocol that gives carriers more capacity on their networks and faster transmission times. Also refers to the FLEX family of protocols: FLEX, InFLEXion and ReFLEX.

Flip Flop. A bistable multivibrator. A circuit which has two output states and is switched from one to the other by means of an external signal (trigger).

Float Charge. Float charge is the voltage required to counteract the self-discharge of the battery at a certain temperature.

Floating Ground. Common connection in a circuit that provides a return path for current but is not connected to an earth ground.

Float Life. Number of years that a battery can keep its stated capacity when it is kept at float charge (see float charge).

Float-zone Process. A method of growing a large-size, high-quality crystal whereby coils heat a polycrystalline ingot placed atop a single-crystal seed. As the coils are slowly raised the molten interface beneath the coils becomes single crystal.

Flow Soldering. Flow or wave soldering technique in large scale electronic assembly to solder all the connections on a printed circuit board by moving the board over a wave of molten solder.

Flutter. Rapid variation of signal parameters, such as amplitude, phase, and frequency. *Note:* Examples of flutter are (a) rapid variations in received signal levels, such as variations that may be caused by atmospheric disturbances, antenna movements in a high wind, or interaction with other signals, (b) in radio propagation, a phenomenon in which nearly all radio signals that are usually reflected by ionospheric layers in or above the E-region experience partial or complete absorption, (c) in radio transmission, rapidly changing signal levels, together with variable multipath time delays, caused by reflection and possible partial absorption of the signal by aircraft flying through the radio beam or common scatter volume, (d) the variation in the transmission characteristics of a loaded telephone circuit caused by the action of telegraph direct currents on the loading coils, (e) in recording and reproducing equipment, the deviation of frequency caused by irregular mechanical motion, *e.g.*, that of capstan angular velocity in a tape transport mechanism, during operation.

Flux. Material used to remove oxide films from the surface of metals in preparation for soldering.

Flux. In magnetism, the magnetic field consisting of lines of force.

Flux Density. The concentration of magnetic lines of force. Determines strength of the magnetic field.

Flywheel Effect. Sustaining effect of oscillation in an *LC* circuit.

FM. *Frequency Modulation.*

FM Improvement Factor. The quotient obtained by dividing the signal-to-noise ratio (SNR) at the output of an FM receiver by the carrier-to-noise ratio (CNR) at the input of the receiver. *Note:* When the FM improvement factor is greater than unity, the improvement in the SNR is always obtained at the expense of an increased bandwidth in the receiver and the transmission path.

FM Improvement Threshold. The point in an FM receiver at which the peaks in the rf signal equal the peaks of the thermal noise generated in the receiver. *Note:* A baseband signal-to-noise ratio of about 30 dB is typical at the improvement threshold, and this ratio improves 1 dB for each decibel of increase in the signal above the threshold.

FM Threshold Effect. In an FM receiver, the effect produced when the desired-signal gain begins to limit the desired signal, and thus noise limiting (suppression). *Note:* FM threshold effect occurs at (and above) the point at which the FM signal-to-noise improvement is measured.

FM Threshold Extension. A change in the value of the FM threshold of a receiver. *Note:* FM threshold extension may be obtained by decreasing the operational bandwidth, thus decreasing the received noise power and allowing the threshold of the desired signal to occur at a lower signal input level.

FNPRM (Further Notice of Proposed Rulemaking). A document issued by the FCC to spur additional comment on a future commission action.

Foliage Attenuation. Reductions in signal strength or quality due to signal absorption by trees or foliage obstructions in the signal's line-of-sight path. For example, 800 MHz systems are seldom deployed in forested areas. Pine needles-nearly the same length as 800 MHz antennas-can negatively affect signal reception in that band.

Forced Intraframe Coding. 1. In motion video compression, immediately following a significant scene change, the differential, predictive and inter- frame coding aspects are temporarily disabled. Then, information is transmitted for the first changed frame, derived entirely from within that frame.

2. In motion video compression, the technique where static parts of the image are periodically refreshed to mitigate the effects of error build-up in the system.

Fortuitous Conductor. Any conductor that may provide an unintended path for signals. *Note:* Examples of fortuitous conductors are water pipes, wires, cables, and metal building and equipment structural members.

Forward Bias. A PN junction bias which allows current to flow through the junction. Forward bias decreases the resistance of the depletion layer.

Forward Echo. In a transmission line, an echo propagating in the same direction as the original wave and consisting of energy reflected back by one discontinuity and then forward again by another discontinuity. *Note:* Forward echoes can be supported by reflections caused by splices or other discontinuities in the transmission medium (*e.g.*, optical fiber, twisted pair, or coaxial tube). In metallic lines, they may be supported by impedance mismatches between the source or load and the characteristic impedance of the transmission medium.

Forward Error Correction. Forward error correction provides correction of data in error for applications where a return signal path is unavailable or system latency is excessive to permit retransmission of the data. It is used for CDs, videoconferencing and DBS, for example.

FEC uses a mathematical technique in which a *syndrome* is generated and transmitted with the data. At the receive location, processing of the syndrome along with the data, allows correction of errors caused by the transmission system. Depending upon the specific *FEC* technique used, and how it is applied, transmitted signals can be made more robust in the presence of transmission errors.

Forward Looking. An action that ignores the past, particularly previous investment. Often used regarding FCC efforts to set interconnection terms; its forward-looking price-setting method ignored phone companies' historic network costs and was struck down by a federal appeals court.

Fossil Fuels. Fuels formed in the ground from the remains of dead plants and animals. It takes millions of years to form fossil fuels. Oil, natural gas, and coal are fossil fuels.

FOT. *Abbreviation for frequency of optimum transmission*. In the transmission of radio waves via ionospheric reflection, the highest effective, *i.e.*, working, frequency that is predicted to be usable for a specified path and time for 90% of the days of the month. *Note 1:* The FOT is normally just below the value of the maximum usable frequency (MUF). In the prediction of usable frequencies, the FOT is commonly taken as 15% below the monthly median value of the MUF for the specified time and path. *Note 2:* The FOT is usually the most effective frequency for ionospheric reflection of radio waves between two specified points on Earth.

FOTS. Fiber Optic Transmission System. With its minimal transit time delay and very low bit error rate, FOTS is the superior method to transmit high-speed data, digital video signals and other compressed signals.

Fourier Analysis. The definition of a periodic waveform of arbitrary shape as a summation of sine waves having specific amplitudes and phases, and having frequencies corresponding to the harmonics of the waveform being defined. *Note:* A Fourier analysis is particularly well suited for communications equipment design and for predicting the performance of a given design. [From Weik =89]

Four-wire Circuit. A two-way circuit using two paths so arranged that the respective signals are transmitted in one direction only by one path and in the other direction by the other path. *Note:* The four-wire circuit gets its name from the fact that, historically, two conductors were used in each of two directions for full-duplex operation. The name may still be applied, *e.g.*, to a communications link supported by optical fibers, even though only one fiber is required for transmission in each direction.

Four-wire Repeater. A repeater, consisting of two amplifiers, one associated with each direction, used in a four-wire circuit.

Four-wire Terminating Set. A balanced transformer used to perform a conversion between 4-wire and 2-wire operation. *Note 1:* For example, a 4-wire circuit may, by means of a 4-

wire terminating set, be connected to a 2-wire telephone set. Also, a pair of 4-wire terminating sets may be used to introduce an intermediate 4-wire loop into a 2-wire circuit, in which loop repeaters may be situated to amplify signals in each direction without positive feedback and oscillation. *Note 2:* Four-wire terminating sets have been largely supplanted by resistance hybrids.

fps. Frames (usually motion video frames) per second.

Frame Delete Compression. In H.261 and H.263 motion video compression, the act of totally discarding from zero to three consecutive video frames is part of the video compression process. Its use depends upon the degree of video motion existing and the aggregate bit rate. Because of the large data loss with this technique, this is truly a compression approach of last resort.

Frame Relay. 1. Wideband, packet-based interface used to transmit bursts of data over a wide-area network. Seldom used for voice. 2. A wide-area network protocol often transported over DS1 facilities. Latency can cause problems for videoconferencing.

Free Electrons. Electrons that are not in any orbit around a nucleus.

Free Running Multivibrator. A multivibrator that produces a continuous output waveform without any signal input. A square wave generator used to produce a clock signal.

Free-space Coupling. Coupling of magnetic, electric, or electromagnetic fields that are not confined to a conductor. *Note:* Coupling by the deliberate introduction of capacitors and inductors is not considered free-space coupling.

F Region. That portion of the ionosphere existing between approximately 160 and 400 km above the surface of the Earth, consisting of layers of increased free-electron density caused by the ionizing effect of solar radiation. *Note 1:* The F region reflects normal-incident frequencies at or below the critical frequency (approximately 10 MHz) and partially absorbs waves of higher frequency. *Note 2:* The F_1 layer exists from about 160 to 250 km above the surface of the Earth and only during daylight hours. Though fairly regular in its characteristics, it is not observable everywhere or on all days. The principal reflecting layer during the summer for paths of 2,000 to 3,500 km is the F_1 layer. The F_1 layer has approximately 5 H 10^5 e/cm^3 (free electrons per cubic centimeter) at noontime and minimum sunspot activity, and increases to roughly 2 H 10^6 e/cm^3 during maximum sunspot activity. The density falls off to below 10^4 e/cm^3 at night. *Note 3:* The F_1 layer merges into the F_2 layer at night. *Note 4:* The F_2 layer exists from about 250 to 400 km above the surface of the Earth. The F_2 layer is the principal reflecting layer for HF communications during both day and night. The horizon-limited distance for one-hop F_2 propagation is usually around 4,000 km. The F_2 layer has about 10^6 e/cm^3. However, variations are usually large, irregular, and particularly pronounced during magnetic storms.

Frequency. Rate of recurrence of a periodic wave. Measured in Hertz (cycles per second).

Frequency Averaging. 1. The process by which the relative phases of precision clocks are compared for the purpose of defining a single time standard. 2. A process in which network synchronization is achieved by use, at all nodes, of oscillators that adjust their frequencies to the average frequency of the digital bit streams received from connected nodes. *Note:* In frequency averaging, all oscillators are assigned equal weight in determining the ultimate network frequency.

Frequency Compatibility. 1. Of an electronic device, the extent to which it will operate at its designed performance level in its intended operational environment (including the presence of interference) without causing interference to other devices. 2. The degree to which an electrical or electronic device or devices operating on or responding to a specified frequency or frequencies is capable of functioning with other such devices.

Frequency Deviation. 1. The amount by which a frequency differs from a prescribed value, such as the amount an oscillator frequency drifts from its nominal frequency. 2. In frequency modulation, the absolute difference between (a) the maximum permissible instantaneous frequency of the modulated wave or the minimum permissible instantaneous frequency of the modulated wave and (b) the carrier frequency. 3. In frequency modulation, the maximum absolute difference, during a specified period, between the instantaneous frequency of the modulated wave and the carrier frequency.

Frequency Dispersal. An electronic counter-countermeasure (ECCM) in which communications nets= operating frequencies are widely separated from each other, causing a requirement to spread jamming power over wider frequency bands and thus compelling a reduction of available jamming power on any single channel or frequency, or causing a requirement for more jamming power or more jamming equipment.

Frequency-division Multiplex (FDM). Transmission of two or more signals over a common path by using a different frequency band for each signal.

Frequency-domain Analysis. A method of representing a waveform by plotting its amplitude against frequency.

Frequency-exchange Signaling. Frequency-change signaling in which the change from one significant condition to another is accompanied by decay in amplitude of one or more frequencies and by buildup in amplitude of one or more other frequencies. *Note:* Frequency-exchange signaling applies to supervisory signaling and user-information transmission.

Frequency Meter. Meter used to measure frequency of periodic waves.

Frequency Multiplier. A harmonic conversion circuit in which the frequency of the output signal is an exact multiple of the input frequency.

Frequency Response. Indication of how well a circuit responds to different frequencies applied to it.

Frequency Response Curve. A graph of amplitude over frequency indicating a circuit response to different frequencies.

Frequency Reuse. The ability of specific channels assigned to a single cell to be used again in another cell, when there is enough distance between the two cells to prevent co-channel interference from affecting service quality. The technique enables a cellular system to increase capacity with a limited number of channels.

Frequency Standard. A stable oscillator used for frequency calibration or reference. *Note 1:* A frequency standard generates a fundamental frequency with a high degree of accuracy and precision. Harmonics of this fundamental frequency are used to provide reference points. *Note 2:* Frequency standards in a network or facility are sometimes administratively designated as "primary" or "secondary." The terms *"primary"* and *"secondary,"* as used in this context, should not be confused with the respective technical meanings of these words in the discipline of precise time and frequency.

Fresnel Lens. An optical device that focuses light like a magnifying glass; concentric rings are faced at slightly different angles so that light falling on any ring is focused to the same point. Fresnel lenses are flat rather than thick in the center and can be stamped out in a mold.

Front Porch. Television usage: In a composite video signal, that portion of the signal between the end of the active video on each horizontal scanning line and the leading edge of the horizontal synchronization pulse.

FRS (Family Radio Service). A very low power, short range two-way radio service in the 460 MHz band.

FSK. *Frequency Shift Keying*, a data transmission modulation method.

FSS. *Fixed Satellite Service*; satellite service to a fixed, non-mobile, location.

Fuel. Any material that can be burned to make energy.

Fuel Cell. A device that converts the energy of a fuel directly to electricity and heat, without combustion. Because there is no combustion, fuel cells give off few emissions; because there are no moving parts, fuel cells are quiet.

Full-Duplex. 1. The radio term applied to transmissions such as telephone calls that allow talking and listening at the same time by using two frequencies to create one channel. Each frequency is used solely for either transmitting or receiving. 2. A transmission system, together with its associated equipment, capable of simultaneously transmitting and receiving signals, as opposed to *simplex* (unidirectional) or *half-duplex* (one direction at a time) systems.

Full Scale Deflection (FDS). Deflection of a meter's pointer to the farthest position on the scale.

.711. An ITU-T recommendation about a 3.4-KHz bandwidth PCM audio coding algorithm working over the range of 32- to 64-Kbps. It forms a part of the ITU-T codec and telephony requirements. This algorithm produces telephone grade audio, especially useful in videoconferencing at aggregate bit rates above 128 Kbps.

Full Wave Rectifier. Rectifier that makes use of the full AC wave in both the positive and negative half cycles.

Function Generator. Signal generator that can produce sine, square, triangle and sawtooth output waveforms.

Fundamental. Of a periodic wave, the sinusoidal component, *i.e.,* Fourier component, having the lowest frequency. *Note:* Every periodic waveform may be expressed as the summation of the fundamental and its harmonics. For example, a square wave may be expressed as the summation of sine waves equal in frequency to the fundamental and all odd harmonics, each frequency having an appropriate amplitude and phase. A pure sinusoidal wave has only one component, *i.e.,* the fundamental.

Fundamental Frequency. Lowest frequency in a complex waveform.

Fuse. 1. A device that has as its critical component a metal wire or strip that will melt when heated by a prescribed (design) amperage, creating an open in the circuit of which it is a part, thereby protecting the circuit from an overcurrent condition. *Note:* Fuses are often characterized as "fast-blow" or "slow-blow," according to the time required for them to respond to an overcurrent condition. Fast-blow fuses open nearly instantaneously when exposed to an overcurrent condition. Slow-blow fuses can tolerate a transient overcurrent condition, but will open if the overcurrent condition is sustained. 2. In optical fiber technology, to join the endfaces of a pair of optical fibers by melting, *i.e.*, welding, the endfaces together.

FWA (Fixed Wireless Access). Also known as wireless local loop.

G

G.720. ITU-T recommendation characterizing low bit rate digital voice codec performance using non-voice signals, useful for testing VoIP coders, transcoders, gateways and terminal devices.

G.722. An ITU-T recommendation detailing a 7-KHz bandwidth audio coding algorithm working at 64-Kbps. It forms a part of the H.320 codec requirements. G.722 produces good quality audio but is generally not recommended at aggregate bit rates below 256 Kbps.

G.722.1. An ITU-T recommendation detailing 7-KHz and 14-KHz bandwidth audio coding algorithms working at 24- and 32-Kbps. It forms part of the H.320 and H.323 codec requirements. May be licensed from Polycom who acquired PictureTel Corp. IP assets. G.722.1 also includes testing and packetizing considerations.

G.723.1. An ITU-T recommendation for a Dual Rate Speech Coder for Multimedia Communications Transmitting at 5.3- and 6.3-Kbps.

G.726. ITU-T recommendation for ADPCM (adaptive differential pulse code modulation) coding at 16-, 24-, 32- and 40-Kbps. Includes useful annexes.

G.727. ITU-T recommendation using ADPCM (adaptive differential pulse code modulation) with 2- to 5-bits/sample. Includes useful annexes with test sequences.

G.728. An ITU-T recommendation detailing a 16-Kbps 3.4-KHz bandwidth audio coding algorithm known as LD-CELP or low-delay code excited linear prediction. Part of the H.320 and H.323 codec recommendations. Produces good speech intelligibility and is a good choice for videoconferencing operating at 112- or 128-Kbps aggregate transmission rates.

G.729. ITU-T recommendation about an 8-Kbps speech coder using the CS-ACELP or, conjugate-structure algebraic-code-excited linear prediction speech coder. Annexes provide algorithms at 6.4- and 11.8-Kbps, C code and test verification suites.

G/T. The figure of merit or carrier-to-noise ratio, of a satellite receive earth station where G is gain and T is noise temperature. The value includes contributions of factors including antenna gain, noise temperature of the low-noise amplifier and other aspects. A typical value for a modern earth station might be 25 db□K.

G. Green. The green video signal, part of an *RGB component* video signal. G may also include synchronization information.

Gain. Increase in voltage, current and/or power. Gain is expressed as a ratio of amplifier output value to the corresponding amplifier input value.

Gain Bandwidth Product. A device parameter that indicates the maximum possible product of gain and bandwidth. The gain bandwidth product of a device is equal to the *unity gain frequency* (f_{unity}) of the device.

Gallium (Ga). A chemical element, atomic number 31, metallic in nature, used in making certain kinds of solar cells and semiconductor devices.

Gallium Arsenide (GaAs). A crystalline, III-V, high-efficiency semiconductor/photovoltaic material.

Gamma Rays. High frequency electromagnetic radiation from radio active particles.

Ganged. Mechanical coupling of two or more capacitors, switches, potentiometers, or any other adjustable components so that adjusting onc control will operate all.

Gas. Any aeriform or completely elastic fluid which is not a solid or a liquid. Gasses are produced by heating a liquid beyond its boiling point.

Gassing Current. Portion of charge current that goes into electrolytical production of hydrogen and oxygen from the electrolytic liquid. This current increases with increasing voltage and temperature.

Gateway. 1. Ground-based link to a mobile satellite service network. 2. Hardware or software that converts between algorithms, protocols or standards. May also act as a means of access from one network to another.

GCC. Generic Conference Control.

Geiger Counter. Device used to detect nuclear particles.

Gel-type Battery. Lead-acid battery in which the electrolyte is composed of a silica gel matrix.

Generator. Device used to convert mechanical energy to electrical energy.

Genlock. The synchronization of a piece of video equipment to another using an external video or similar signal. In broadcasting and videoconferencing systems, all picture sources should be genlocked together or made *synchronous*.

Geostationary Orbit Satellite System. A communications system with satellites in geosynchronous orbits 22,300 miles above the Earth. These satellites appear stationary because they move at the same rate as the Earth's rotation.

Geosynchronous. Maintaining a fixed orbit, about 24,000 miles above the Earth.

GFSK. *Gaussian filtered Frequency Shift Keying*; a data modulation method used by Bluetooth (which see) wireless systems. Gaussian filtering reduces the instantaneous transmission bandwidth required.

Giga. Metric prefix for 1 billion (10^9).

Gigahertz (GHz). One billion radio waves, or cycles, per second. Equal to one thousand Megahertz.

Gigawatt (GW). One billion watts. One million kilowatts. One thousand megawatts.

Glazings. Clear materials (such as glass or plastic) that allow sunlight to pass into solar collectors and solar buildings, trapping heat inside.

GLONASS (Global Navigation System). A Russian satellite location technology similar to global positioning system.

GMPCS (Global Mobile Personal Communications Services). A term that refers to future mobile satellite systems that will provide wireless phone service anywhere in the world.

GPRS (General Packet Radio Services). a 2.5-generation technology (being implemented in GSM networks) that may offer wireless data access speeds of up to 144 kilobits per second in end-user devices.

GPS (Global Positioning System). A series of 24 geosynchronous satellites that continuously transmit their position. Used in personal tracking, navigation and automatic vehicle location technologies.

Grain Boundaries. The boundaries where crystallites in a multicrystalline material meet.

Gray Market. May refer to a place where a carrier bought a product from another carrier, not direct from a vendor or the vendor's distribution channel. The exact location of the seller/buyer is unknown to the manufacturer.

Greenhouse Effect. The effect of the Earth's atmosphere, due to certain gases, in trapping heat from the sun; the atmosphere acts like a greenhouse.

Greenhouse Gases. Gases that trap the heat of the sun in the Earth's atmosphere, producing the greenhouse effect; the two major greenhouse gases are water vapor and carbon dioxide; lesser greenhouse gases include methane, ozone, chlorofluorocarbons, and nitrogen oxides.

Grid. See 'Electrical grid'.

Grid-connected (PV system). A PV system in which the PV array acts like a central generating plant, supplying power to the grid.

Grid-interactive (PV system). See 'Grid-connected (PV system)'.

Ground. An intentional or accidental conducting path between an electrical system or circuit and the earth or some conducting body acting in place of the earth. A ground is often used as the common wiring point or reference in a circuit.

Ground Loop. An undesirable feedback condition caused by two or more circuits sharing a common electrical line, usually a grounded conductor.

GSM (Global System for Mobile Communications). A digital cellular or PCS network used throughout the world.

GSM-Plus. An enhanced version of global system for mobile communications technology that will be developed to meet IMT-2000 capabilities.

GSM-R. Global system for mobile communications for railway networks. GSM-R uses standard base station and switching infrastructure to provide fast data transmission for railways.

GSO. *GeoStationary Orbit*; a satellite in an equatorial orbit and at an altitude that matches the rotation of the Earth. To a fixed observer on the Earth, the satellite position appears stationary.

GUI (Graphical User Interface). 1. A computing term referring to an operating system or environment that displays options on the screen as graphical symbols, icons or photographs.

2. *Graphical User Interface*; a human - machine interface based on a graphical and iconic metaphor techniques, to simplify or clarify operation of a device and to provide visual feedback to the operator about the state of the system.

Gunn Diode. A semiconductor diode that utilizes the Gunn effect to produce microwave frequency oscillation or to amplify a microwave frequency signal.

H

H. Horizontal.

H.221. ITU-T recommendation regarding frame structure for audiovisual teleservices operating between 64 kbps and 1920 kbps.

H.222.1. ITU-T recommendation regarding multimedia multiplex and synchronization for audio-visual communication in an ATM environment.

H.223. ITU-T recommendation describing a multiplexing protocol for low bit rate multimedia cCommunication. Annexes include considerations for mobile and other high error rate environments.

H.224. ITU-T recommendation regarding a real-time control protocol for simplex operation using H.221 LSD, HSD and MLP channels.

H.230. ITU-T codec recommendation regarding frame-synchronous control and indication signals for audio-visual systems.

H.231. ITU-T MCU recommendation for multipoint control units using audio-visual channels up to 1920 Kbps.

H.233. ITU-T recommendation regarding a confidentiality system for audio-visual services.

H.234. ITU-T recommendation regarding encryption key management and authentication system for audiovisual teleservices.

H.235. ITU-T recommendation regarding security and encryption for H.323 and other H.245 based multimedia terminals.

H.241. ITU-T recommendation regarding extended procedures and control signals for H.300-series terminals.

H.242. ITU-T codec recommendation regarding a system for establishing communication between audio-visual terminals using digital channels up to 2 Mbps.

H.243. ITU-T recommendation defining procedures for establishing communication between three or more audio-visual terminals using digital channels of up to 1920 Kbps.

H.244. ITU-T recommendation regarding synchronized aggregation of multiple 64 or 56 Kbps channels (a.k.a. BONDING).

H.245. ITU-T recommendation setting out a control protocol for multimedia communication.

H.248.x. A series of 35 ITU-T recommendations regarding various forms of gateway control protocols.

H.261. ITU-T recommendation regarding a video codec for audio-visual services at p x 64 Kbps. Note that motion compensation and field sub-sampling (loop filter) are implementation-optional in the standard. Users considering the H.261 codec at bit rates of 384 Kbps or below should verify that the manufacturer has implemented these optional aspects.

H.261 Annex D. ITU-T codec recommendation for simultaneous graphics transmission in the North American region. Other administrations may use *JPEG*.

H.263. ITU-T recommendation regarding a video codec for audio-visual services, video coding for low bit rate communication. Considerably broadens H.261 capabilities, extending from low- resolution, credit card-size images, to high-quality, exceptionally detailed images for applications such as telemedicine. Backward compatibility is provided with H.320 QCIF (Quarter Common Intermediate Format) and optionally to CIF and SQCIF (Sub-QCIF).

The following table compares H.320 and H.324 video capabilities:

Picture Format	*Luminance Pixels*	*Video Modes* *H.320/H.261*	*H.324/H.263*
SQCIF	128H x 96V	O	R
QCIF	176H x 144V	R	R
CIF	352H x 288V	O	O
4CIF	704H x 576V	-	O
16CIF	1408H x 1152V	-	O

Where: S = Sub, Q = Quarter, H = Horizontal pixels, V = Vertical pixels, R = Required, O = Optional and, - = Not Defined.

H.262. ITU-T recommendation regarding advanced video coding for high-quality applications. This standard is the result of the merging of the ISO/IEC MPEG2 (which see) work and the ITU-T H.262.

H.264. ITU-T recommendation regarding advanced video coding for high-quality motion video applications. This standard is the result of the merging of the ISO/IEC MPEG4 Part 10 AVC and the ITU-T H.26L committees, to avoid duplication of parallel development efforts. In general, H.264 provides quality superior to MPEG2 at about 30 per cent of the bit rate or less. It is expected to fully replace MPEG2 over time.

There are two significant versions of H.264: Level 3 provides for mainstream video while Level 4 focuses on HDTV. Significant changes from MPEG2 include 4x4 PEL block-size and what is called foreground-background video coding.

H.281. ITU-T recommendation regarding far-end camera control protocol for videoconferencing using H.224.

H.282. ITU-T recommendation regarding remote device control protocol for multimedia applications.

H.310. ITU-T recommendation regarding broadband audio-visual terminals and communication systems.

H.320 (suite spec.). ITU-T codec recommendation bringing together other recommendations to produce an overall visual telephony system specification including terminal equipment operating over a narrow bandwidth between 56 Kbps and 1920 Kbps.

H.321. ITU-T recommendation describing adaptation of H.320 to an ISDN *n* times ISDN B-channel environment.

H.322. ITU-T recommendation describing visual telephone systems and terminal equipment for local area networks that provide a guaranteed quality of service (guaranteed bandwidth). This standard makes direct reference to the Draft IEEE *P802.9a*, (which see). These standards describe low latency LANs, suitable for videoconferencing (Iso-Ethernet).

H.323 (suite spec.). ITU-T recommendation describing packet-based visual telephone systems and terminal equipment for local area networks that provide a non-guaranteed quality of service. Includes multiple annexes.

H.324 (suite spec.). ITU-T recommendation describing a terminal for low bit rate multimedia communication; includes various recommendations including video codec for low-speed connections using single or dual V.34 connected to analog POTS, plus ISDN, ATM and probably mobile, bringing together other recommendations to produce an overall visual telephony system specification. Some compatibility with H.320, network transcoding, exceptionally good multipoint support and continuous presence are also provided.

Half-Duplex. The radio term applied to transmissions which allow two-way communications over a single frequency through the use of a push-to-talk button that opens and closes the communication pathway over that frequency.

Half Power Point. A frequency at which the power is 50% of maximum. This corresponds to 70.7% of maximum current or voltage.

Half Wave Rectifier. A diode rectifier that converts AC to pulsating DC by eliminating either the negative or the positive alternation of each input AC cycle.

Handheld Personal Computer. A palmtop device or personal digital assistant that can wirelessly access e-mail, paging messages and the Internet/intranets.

Handoff. The process occurring when a wireless network automatically switches a mobile call to an adjacent cell site.

Hands-free. A feature for mobile phones that allows the driver to use their car phone without lifting or holding the handset to their ear.

H-parameters (hybrid parameters). Transistor specifications that describe the component operating limits under specific circumstances.

Harmonic. Sine wave that is smaller in amplitude and some multiple of a fundamental frequency. Example: 880 Hz. is the second harmonic of 440 Hz., 880 Hz. is the third harmonic of 220 Hz.

Hartley Oscillator. An oscillator that uses a tapped inductor in the feedback network.

HDB3. *High Density Bipolar 3.* Zero suppression scheme used for signals transmitted over E1 links, similar to B8ZS and is based on AMI. A European digital network transmission line protocol allowing information and control data to both be embedded in the transmission bit stream.

HDLC. ISO: High-level Data Link Control protocol.

HDSL. *High bit-rate Digital Subscriber Line,* requiring two pairs, it can provide full duplex 1.544 Mbps over a distance of 3.65 Km (2.25 miles), mimicking a T1 line. This type of DSL does not support simultaneous telephony over the same physical wires. HDSL is a good choice for applications like videoconferencing requiring symmetrical communication links.

HDTV (High Definition Television). 1. Digital television signals transmitted in the very high frequency band by national and local TV stations. HDTV promises improved image broadcasts and compact disc-quality sound.

2. Any of a number of television technologies providing an image aspect ratio of about 2:1 (WxH) and with horizontal and vertical resolution far superior to NTSC, PAL or SECAM.

Heat Pump. Like an air conditioner or refrigerator, a heat pump moves heat from one location to another. In the cooling mode, heat pumps reduce indoor temperatures in the summer by transferring heat to the ground. Unlike an air conditioning unit, however, a heat pump's cycle is reversible. In winter, a heat pump can extract heat from the ground and transfer it inside. The energy value of the heat thus moved can be more than three times the cost of the electricity required to perform the transfer process.

Henry. The basic unit of inductance.

HERTZ (Hz.). 1. Unit of frequency. One hertz is equal to one cycle per second. 2. A unit of measurement of one cycle per second, or one radio wave passing one point in one second of time. Named in honor of Heinrich Hertz, the discoverer of the theory of radio waves.

Heterojunction. A region of electrical contact between two different semiconductor materials.

High Fidelity (Hi Fi). Sound reproduction equipment that reproduces sound as near to the original sound as possible

High-pass Filter. A tuned circuit designed to pass all frequencies above a designated cut-off frequency. Frequencies below the cut-off frequency are rejected or attenuated

High Tension. Lethal voltage in the kilovolt range and above.

High-yield or Junk Financing. Bonds or other debts incurred by companies considered speculative or below 'investment grade' by major credit rating agencies such as Standard & Poor's, Moody's Investors Service, Duff & Phelps or Fitch. Because a lender to such companies takes on a greater risk that the borrower will default-a typical risk with startup companies in PCS or financially stressed companies in paging-the borrower will pay a higher interest rate or yield on the loan.

HLR (Home Location Register). A database residing in a local wireless network that checks the identity of a local subscriber.

Hole. 1. A gap left in the covalent bond when a valence electron gains sufficient energy to jump to the conduction band. 2. The vacancy where an electron would normally exist in a solid; behaves like a positively charged particle.

Hologram. Three-dimensional picture created with a laser.

Holography. The science dealing with three-dimensional optical recording.

Homojunction. The region between an n-layer and a p-layer in a single material photovoltaic cell.

Horizontally Polarized Wave. Electromagnetic wave that has the electric field in the horizontal plane.

Hot Spot. An undesirable phenomenon of PV device operation whereby one or more cells within a PV module or array act as a resistive load, resulting in local overheating or melting of the cell(s).

HTML. *HyperText Markup Language,* a subset of SGML, commonly used for Internet Web-page design.

HTTP. *HyperText Transport Protocol,* the method used for Internet domain addressing.

Huffman Coding. A lossless, statistically-based entropy coding technique used to compress data in which the most frequently occurring code groups are represented by shorter codes, and rarely occurring code groups are represented by longer codes. Used in H.320 and other videoconferencing codecs, Group 3 facsimile and JPEG.

Hybrid Circuit. Circuit that combines two technologies (passive and active or discrete and integrated components) onto one microelectronic circuit. Passive components are usually made by thin film techniques, while active components are made with semiconductor techniques.

Hybrid System. A PV system that includes other sources of electricity generation, such as wind or diesel generators.

Hydrogenated Amorphous Silicon. Amorphous silicon with a small amount of incorporated hydrogen. The hydrogen neutralizes dangling bonds in the amorphous silicon, allowing charge carriers to flow more freely.

Hysteresis. Amount that the magnetization of a material lags the magnetizing force due to molecular friction. In Schmitt Trigger circuits, the difference between the upper and lower trigger points.

Hz. *Hertz,* analog usage, cycles per second.

I

IC. Abbreviation for "integrated circuit".

IC Voltage Regulator. Three terminal device used to hold the output voltage of a power supply constant over a wide range of load variations.

iDEN (Integrated Digital Enhanced Network). A Motorola Inc. enhanced specialized mobile radio network technology that combines two-way radio, telephone, text messaging and data transmission into one network.

IDSL. *ISDN-like Digital Subscriber Line,* the slowest form of xDSL, it provides full duplex symmetrical connectivity at 144 Kbps over distances up to 10.7 Km (6.6 miles). For many rural users, it may be the only higher-speed alternative over V.90 or slower modems. This type of transmission does not support telephony over the same physical pair as does ADSL.

IEE. Institution of Electrical Engineers (London, UK).

IEEE. Institute of Electrical and Electronics Engineers (New York City, USA) (formerly IRE, Institute of Radio Engineers).

IGFET. Insulated gate field effect transistor. Another name for a "MOSFET."

ILEC (Incumbent Local Exchange Carrier). The historic local phone service provider in a market, often a former Bell company. Distinct from CLECs, competitive local exchange carriers, new market entrants.

Immunity. Immunity has special meaning in a 911 context. No CMRS or 911 provider, its employees, officers or agents is criminally liable or liable for any damages in a civil action for injuries, death or loss to person or property resulting from any act or omission in connection with the development, adoption, implementation, maintenance, enhancement or operation of E911 service, unless such damage or injury was intentional or the result of gross negligence or willful or wanton conduct.

Impedance (Z). Measured in ohms it is the total opposition to the flow of current offered by a circuit. Impedance consists of the vector sum of resistance and reactance.

Impedance Coupling. Coupling of two signal amplifier circuits through the use of an impedance such as a inductor.

Impedance Matching. Matching the output impedance of a source to the input impedance of a load to attain maximum power transfer.

Improved Mobile Telephone Service (IMTS). The commercial form of mobile telephone service preceding cellular that allowed users to place and receive their own calls through the use of a dial or keypad on the telephone. Prior service (MTS) required an operator's intervention to actually place or receive a call.

IMT-2000. The International Telecommunication Union's name for the new third generation global standard for mobile telecommunications.

IMTA (International Mobile Telecommunications Association). A trade group serving specialized mobile radio and public access mobile radio carriers around the world.

Incandescence. State of a material when heated to the point where it emits light. (red hot or white hot).

Incident Light. Light that shines onto the face of a solar cell or module.

Indium Oxide. A wide band gap semiconductor that can be heavily doped with tin to make a highly conductive, transparent thin film. Often used as a front contact or one component of a heterojunction solar cell.

Induced Voltage. Voltage generated in a conductor when subjected to a moving magnetic field.

Inductance. Property of a circuit to oppose a change in current. The moving magnetic field produced by a change in current causes an induced voltage to oppose the original change.

Inductive Circuit. Circuit having greater inductive reactance than capacitive reactance.

Inductive Reactance. Opposition to the flow of AC current produced by an inductor. Measured in Ohms and varies in direct proportion to frequency.

Inductor. Length of conductor used to introduce inductance into a circuit. The conductor is usually wound into a coil to concentrate the magnetic lines of force and maximize the inductance. While any conductor has inductance, in common usage the term inductor usually refers to a coil.

InFLEXion. The narrowband PCS technology developed by Motorola Inc. that allows for voice paging. It has been adopted by carriers such as Paging Network Inc. and Conxus Communications Inc.

Infrared. Electromagnetic heat radiation whose frequencies are above the microwave frequency band and below red in the visible band.

Infrared Radiation. Electromagnetic radiation whose wavelengths lie in the range from 0.75 micrometer to 1000 micrometers.

Inhibit. To stop an action or block data from passing.

In Phase. When two or more waves of the same frequency have their positive and negative peaks occurring at the same time.

Input Impedance. Opposition to the flow of signal current at the input of a circuit or load.

Insolation. Sunlight, direct or diffuse; from 'incident solar radiation'. Not to be confused with 'insulation'. Equal to about 1000 watts per square meter at high noon.

Insulated. When a non conducting material is used to isolate conducting materials from one another.

Insulation. Materials that prevent or slow down the movement of heat.

Insulating Material. Material that will prevent the flow of current due to its chemical composition.

Insulation Resistance. Resistance of insulating material. The greater the insulation resistance, the better the insulation.

Integrated. When two or more components are combined into a circuit and then incorporated into a single package.

Integrator. A device that approximates and whose output is proportional to an integral of the input signal. A low pass filter.

Interconnect. A conductor within a module or other means of connection which provides an electrical interconnection between the solar cells. [UL 1703]

Interconnection. The connecting of one network with another, e.g. a cellular carrier's wireless network with the local exchange.

Interconnection Charge. A fee charged by the local telephone operating company to the wireless operator for completing a call into the landline network (the PSTN).

Intermediate Frequency Amplifier. In a superheterodyne radio it amplifies a fixed frequency lower than the received radio frequency and higher than the audio frequency.

Intermittent. A fault occurring at random intervals of time. Intermittent problems are often difficult to locate because of the random nature. They often don't occur when the technician is present.

Internal Resistance. Every source has some resistance in series with the output current. When current is drawn from the source some power is lost due to the voltage drop across the internal resistance. Usually called output impedance or output resistance.

Interoperability. 1. The ability of a network to operate with other networks, such as two systems based on different protocols or technologies.

2. The degree of compatibility among devices required to operate with one another with reference to a specified feature set or standard. This is an important consideration when mixing non-ITU-T compliant systems from different manufacturers or when mixing ITU-T compliant systems from different manufacturers which often may support different capabilities, allowable within the standards. Remember that many ITU-T systems revert to the highest common denominator for interability.

Intrinsic Material. A semiconductor material with electrical properties essentially characteristic of ideal pure crystal. Essentially silicon or germanium crystal with no measurable impurities.

Intrinsic Semiconductor. An undoped semiconductor.

Intrinsic Stand-off Ratio. A unijunction transistor (UJT) rating used to determine the firing potential of the device.

Inverters. Devices that convert dc electricity into ac electricity (single or multiphase), either for stand-alone systems (not connected to the grid) or for utility-interactive systems.

Inverting Amplifier. An amplifier that has a 180□ phase shift from input to output.

Inverting Input. In an operational amplifier (op amp) the input that is marked with a minus sign. A signal applied at the inverting input will be given 180 phase shift between input and output.

Ion. An atom with fewer electrons in orbit than the number of protons in the nucleus is a positive ion. An atom with a greater number of electrons in orbit than the number of protons in the nucleus is a negative ion.

Ionized. Atoms become ionized when they gain or lose a valence electron.

IP. Internet Protocol, packet transmission standard for the transmission of data, voice, video and other information over the Internet.

IP PBX. IP Private Branch Exchange. A telephone, data and video switching system, usually located on customer premises and belonging to the user. Makes use of technologies like VoIP, H.323 and Ethernet.

IR. *InfraRed*, that portion of the electromagnetic spectrum adjacent to the visible light spectrum and with longer wavelengths. IR is often used for optical communications over FOTS or through the air.

IRD. *Integrated Receiver-Decoder*, the device to receive and decode television signals from DTH (direct to home) broadcast satellites or MMDS (wireless cable) systems.

IRE Scale Units. Institute of Radio Engineers scale units. A standard measure of NTSC video signal amplitudes for which 140 IRE scale units equals 1.00 volt peak-to-peak.

IS (Interim Standard). A designation of the American National Standards Institute—usually followed by a number—that refers to an accepted industry protocol; e.g, IS-95, IS-136, IS-54.

IS-136 (Interim Standard). The latest generation of the digital standard time division multiple access technology.

IS-41 (Interim Standard). The network standard that allows all switches to exchange information about subscribers.

IS-54 (Interim Standard). The first generation of the digital standard time division multiple access technology.

IS-661 (Interim Standard). North American standard for 1.9 GHz wireless spread spectrum radio-frequency access technology developed by Omnipoint Corp. IS-661, for which Omnipoint was awarded a pioneer's preference license for the New York City market, is based on a composite of code division multiple access and time division multiple access technologies. The company says IS-661 reduces infrastructure costs and allows higher data speeds than mainstream GSM or TDMA platforms.

IS-95 (Interim Standard). The standard for code division multiple access.

ISDN (Integrated Services Digital Network). 1. An advanced, high-capacity wireline technology used for high-speed data transfer. 2. A switched digital service using a 16 Kbps "D" channel for signaling and two or more 64 Kbps or 56 Kbps (some US) "B" channels for data transmission.

ISM. *Industrial, Scientific and Medical*, unlicensed radio bands.

ISO. International Standards Organization (Geneva, CH).

ISPRA Guidelines. Guidelines for the assessment of PV Plants, published by the Joint Research Centre of the Commission of the European Communities, Ispra, Italy.

ITA (Industrial Telecommunications Association). A Washington, D.C. trade group serving private wireless licensees such as airlines and oil companies.

ITU (International Telecommunication Union). An agency of the United Nations, headquartered in Geneva, that furthers the development of telecommunications services worldwide and oversees global allocation of spectrum for future uses.

ITU-R (formerly CCIR). An International Telecommunications Union (United Nations) international telecommunications standards-setting body, specializing in radio-spectrum usage.

ITU-T (formerly CCITT). An International Telecommunications Union (United Nations) international telecommunications standards-setting body specializing in telecommunications.

ITU-T-Compatible Codec. *jargon:* For example, a video codec which meets all the minimum requirements as set out in ITU-T Recommendations H.221, H.230, H.242, H.261 and H.320. To be compatible with the ITU-T Multipoint Control Unit (MCU), codecs also need to generate and receive the commands in H.231 and H.243. T.120 support is also desirable.

I-type Semiconductor. Semiconductor material that is left intrinsic, or undoped so that the concentration of charge carriers is characteristic of the material itself rather than of added impurities.

I-V Curve. A graphical presentation of the current versus the voltage from a photovoltaic device as the load is increased from the short circuit (no load) condition to the open circuit (maximum voltage) condition. The shape of the curve characterized cell performance.

I-V Data. The relationship between current and voltage of a photovoltaic device in the power-producing quadrant, as a set of ordered pairs of current and voltage readings in a table, or as a curve plotted in a suitable coordinate system (i.e., Cartesian). [ASTM E 1036]

IXC. Inter-eXchange Carrier.

J

J. A prefix used to indicate an imaginary number. (Operator j)

Jack. Socket or connector into which a plug may be inserted.

Java 2 Micro Edition (aka J2ME). Java 2 Micro Edition is a mobile platform enabling carriers to offer subscribers downloadable applications on wireless phones. Developed by Sun Microsystems, J2ME enables developers to write a single application program that is compatible with several types of devices. The platform directly competes with BREW — Binary Runtime Environment for Wireless — backed by Qualcomm.

JFET. Abbreviation for "Junction field effect transistor".

JPEG. Joint Photographic Experts Group. An *ISO* video compression standard for storage and transmission of a wide variety of still graphics image formats. Currently used by digital cameras and is commonly used for graphics on the Internet.

In telecommunications and videoconferencing, JPEG may be used in conjunction with fully compliant ITU-T codecs and includes both lossy and lossless modes. Note that so-called "motion JPEG" is a proprietary method of motion video transmission, and is not a standard.

Joule. The unit of work and energy.

Junction. 1. Contact or connection between two or more wires or cables. The area where the *p*-type material and *n*-type material meet in a semiconductor. 2. A region of transition between semiconductor layers, such as a p/n junction, which goes from a region that has a high concentration of acceptors (p-type) to one that has a high concentration of donors (n-type).

Junction Box. A PV generator junction box is an enclosure on the module where PV strings are electrically connected and where protection devices can be located, if necessary.

Junction Diode. 1. diode. A semiconductor diode in which the rectifying characteristics occur at a junction between the *n*-type and *p*-type semiconductor materials. 2. A semiconductor device with a junction and a built-in potential that passes current better in one direction than the other. All solar cells are junction diodes.

K

Ka-Band. Radio spectrum in the 18 GHz to 31 GHz range used by satellite communications systems.

Kbps (Kb/s). Kilobits (thousands of bits) per second. One kilobit contains 1,024 bits (2^{10} bits)

KBps (KB/s). KiloBytes (thousands of bytes or octets) per second. One kilobyte contains 1,024 bytes (2^{10} bytes)

Key. A sequence of symbols applied to data in order to provide encryption. Sometimes referred to as the "seed".

KG. Key Generator. A device external to a codec or other telecommunications equipment for maintaining the security of classified communications using encryption techniques. Provides much better security than *DES* and is generally used only by the military, NASA, NATO and some military contractors.

KHz. *KiloHertz,* (analog), thousands of cycles per second.

Kilo. Metric prefix for 1000 (10^3).

Kilohertz (KHz). One thousand radio waves, or cycles, per second.

Kilovolt-ampere. 1000 volts at 1 ampere.

Kilowatt (kW). 1000 watts.

Kilowatt-hour (kWh). One thousand watts acting over a period of 1 hour. The kWh is a unit of energy. 1 kWh=3600 kJ.

Kilowatt-hour Meter. A meter used by electric utility companies to measure the amount of electric power used by a customer.

Kinetic Energy. Energy associated with motion.

Kirchhoff's Current Law. The sum of the currents flowing into a point in a circuit is equal to the sum of the currents flowing out of that same point.

Kirchhoff's Voltage Law. The algebraic sum of the voltage drops in a closed path circuit is equal to the algebraic sum of the source voltages applied.

Knee Voltage. The voltage at which a curve joins two relatively straight portions of a characteristic curve. For a PN junction diode, the point in the forward operating region of the characteristic curve where conduction starts to increase rapidly. For a zener diode, the term is often used in reference to the zener voltage rating.

Ku-Band. 1. Radio spectrum in the 10.9 GHz to 17 GHz range used by satellite communications systems.

2. A microwave frequency band. When used for satellite links, the frequencies assigned are:

— uplink (to the satellite): 14.0-14.5 GHz
— downlink (from the satellite): 11.7-12.2 GHz

Ku-band is susceptible to rainfall attenuation, often noted by DBS customers as blocking.

L

Lag. Difference in time between two waveforms of the same frequency expressed in degrees. Example: One waveform lags another waveform by a certain number of degrees.

Laminated Core. Core made up of sheets of magnetic material insulated from one another by an oxide or varnish.

Lamp. Device that produces light.

LAN (Local Area Network). A group of client computers connected to a server.

LAP. Link Access Protocol (frame relay use, LAPD).

Laser. Device that produces a very narrow intense beam of light. The name is an acronym for "light amplification by stimulated emission of radiation.

LATA. Local Access and Transport Area, a telecommunications tariff aspect in the USA.

Latency. The amount of delay in a LAN or WAN. For basic data where a small delay can be tolerated, latency is usually not an issue. However, for communications services used for videoconferencing or VoIP for example, latency can interfere with the audio and/or visual communications. In shared bandwidth transmission environments, it is possible to encounter latency which varies dynamically, caused by perhaps a single user accessing or originating multi-megabyte-sized files or accessing high bandwidth streaming signals.

Lattice. The regular periodic arrangement of atoms or molecules in a crystal.

LBR. Low bit rate.

LCD. Liquid Crystal Display. A very low-power device capable of displaying characters, words and symbols, often built into a telecommunications equipment control panels.

LCP. Link Control Protocol (PPP usage).

LCR. Least-Cost Routing.

LDAP. *Lightweight Directory Access Protocol*, developed at the University of Michigan as a front-end to the more complex ITU-T X.500 directory protocol, formalized in 1995 in RFC 1777. It includes five models: data, organization, security, function and topology. It is useful for a wide range of uses including simple dialing directories and indexing large databases.

Lead. The angle by which one alternating signal leads another in time. Opposite of lag. Also a wire that connects two points in a circuit.

Lead-acid Cell. Cell made up of lead plates immersed in a sulphuric acid electrolyte. An automobile battery usually consists of six lead-acid cells.

Leakage. Small undesireable flow of current through an insulator or dialectric.

LEC (Local Exchange Carrier). A wireline phone company serving a local area.

LED. Abbreviation for "light emitting diode."

Left-hand-rule. If fingers of the left hand are placed around a wire so that the thumb points in the direction of electron flow, the fingers will be pointing in the direction of the magnetic field being produced by the conductor.

Lenz's Law. The current induced in a circuit due to a change in the magnetic field is so directed as to oppose the flux, or to exert a mechanical force to oppose the motion.

LEO (Low-earth Orbit). A mobile communications satellite between 700 and 2,000 kilometers above the earth.

Level Detector. An op-amp circuit that compares two inputs and provides a DC output indicating the polarity relationship between the inputs. A comparitor.

Lie Detector. Piece of electronic equipment also called a polygraph used to determine whether a person is telling the truth by looking for dramatic changes in blood pressure, body temperature, breathing rate, heart rate and skin moisture in response to questions.

LifePage. A service of the PCIA Foundation that donates pagers to individuals awaiting organ donor transplants.

Lifetime. The time from the creation of an electron hole pair until recombination occurs.

Light. Electromagnetic radiation in a band of frequencies that can be received by the human eye.

Light-emitting Diode. A semiconductor diode that converts electric energy into electromagnetic radiation at a visible and near infrared frequencies when its *pn* junction is forward biased.

Light-induced Defects. Defects, such as dangling bonds, induced in an amorphous silicon semiconductor upon initial exposure to light.

Light Trapping. The trapping of light inside a semiconductor material by refracting and reflecting the light at critical angles; trapped light will travel further in the material, greatly increasing the probability of absorption and hence of producing charge carriers.

Limiter. Circuit or device that prevents some portion of its input from reaching the output. A clipper.

Linear. Relationship between input and output in which the output varies in direct proportion to the input.

Linear Scale. A scale in which the divisions are uniformly spaced.

Line-commutated Inverter. An inverter that is tied to a power grid or line. The commutation of power (conversion from dc to ac) is controlled by the power line, so that, if there is a failure in the power grid, the PV system cannot feed power into the line.

Line Regulation. The ability of a voltage regulator to maintain a constant voltage when the regulator input voltage varies.

Lip Sync. The maintenance of sound (i.e., speech) exactly in step with movement in a visual image (*e.g.*, talking heads). A common cause is different transmission paths and transmission technologies for the sound and image signals.

Within codec hardware, the processing time for the video portion of the signal is about 100 times longer than the audio processing time. Consequently, codecs usually incorporate adjustable audio delay circuitry to delay-equalize the two signals to maintain lip sync.

Lithium Ion. The latest form of battery technology, allowing the highest power-to-weight ratio (lots of power for very little weight) and no "memory effect," which plagues batteries made from other materials.

Little LEO. A low-earth orbiting satellite system primarily providing data services; e.g., Leo One, Orbcomm.

Live. Term used to describe a circuit or piece of equipment that is on and has current flow within it.

LMCC (Land Mobile Communications Council). A trade group of frequency coordinators and associations serving private users and commercial operators.

LMCS. *Local Multipoint Communication System*, another version of MMDS or "wireless cable TV"; a term used in Canada. Data rates can be as high as 155 Mbps compared to 64 Kbps cell phones and, depending on jurisdiction, may operate between 2-GHz and 42-GHz with up to 3-GHz of spectrum available.

LMDS (Local Multipoint Distribution Service). Located in the 28 GHz and 31 GHz bands, LMDS is a broadband radio service designed to provide two-way transmission of voice, high-speed data and video (wireless cable TV). FCC rules prohibit incumbent local exchange carriers and cable TV companies from offering in-region LMDS.

LMR. *Land Mobile Radio.*

LMSS. *Land Mobile Satellite Service.*

LNA. *Low Noise Amplifier*, an active device usually used in conjunction with an appropriate antenna to amplify and receive satellite microwave signals.

LNB or LNBC. *Low Noise Block Converter*, an active device used in conjunction with an appropriate antenna to convert a range of satellite signals to some intermediate frequency for processing by a receiver etc.

LNP (Local Number Portability). The ability of subscribers to switch local or wireless carriers and still retain the same phone number, as they can now with long-distance carriers. Wireless carriers don't have to offer LNP until March 2000 and want the deadline further postponed.

Load. 1. A source drives a load. Whatever component or piece of equipment is connected to a source and draws current from a source is a load on that source. 2. Anything in an electrical circuit that, when the circuit is turned on, draws power from that circuit.

Load Current. Current drawn from a source by a load.

Load Impedance. Vector sum of reactance and resistance in a load.

Loading Effect. Large load impedance will draw a small load current and so loading of the source is small. (light load). A small load impedance will draw a large load current from the source. (heavy load).

Load Regulation. The ability of a voltage regulator to maintain a constant output voltage under varying load currents.

Load Resistance. Resistance of a load.

Local Calling Area. The region across which the call is truly local, involving no toll charges.

Logic. Science of dealing with the principle and applications of gates, relays and switches.

Loop Filter. H.261, H.263 video compression algorithm usage. See *Spatial Filtering.*

Loss. Term used to describe a decrease in power.

Low Pass Filter. A tuned circuit designed to pass all frequencies below a designated cut-off frequency.

LSGAC (Local-State Governmental Advisory Committee). An FCC-established group that is working on an antenna-siting solution. The LSGAC will advise carriers and communities on antenna siting.

Luminance. The portion of a composite video signal that represents the monochrome or brightness part of the image.

LZW. Lempel-Ziv-Welch; a lossless data compression algorithm named after its developers.

M

Magnet. Body that can be used to attract or repel magnetic materials.

Magnetic Circuit Breaker. Circuit breaker that is tripped or activated by use of an electromagnet.

Magnetic Coil. Spiral of a conductor which is called an electromagnet.

Magnetic Core. Material that exists in the center of the magnetic coil to either physically support the windings (non-magnetic material) or to concentrate the magnetic flux (magnetic material).

Magnetic Field. Magnetic lines of force travelling from the north pole to the south pole of a magnet.

Magnetic Flux. The magnetic lines of force produced by a magnet.

Magnetic Leakage. The passage of magnetic flux outside the path along which it can do useful work.

Magnetic Poles. Points of a magnet from which magnetic lines of force leave (north pole) and arrive (south pole).

Magnetism. Property of some materials to attract or repel others.

Magnetizing Force. Also called magnetic field strength. It is the magnetomotive force per unit length at any given point in a magnetic circuit.

Magnetomotive Force. Force that produces a magnetic field.

Majority Carriers. 1. The conduction band electrons in an n-type material and the valence band holes in a p-type material. Produced by pentavalent impurities in n-type material and trivalent impurities in p-type material.

2. Current carriers (either free electrons or holes) that are in excess in a specific layer of a semiconductor material (electrons in the n-layer, holes in the p-layer) of a cell.

Marginal Cost. The cost of one additional unit within a group of like units.

Matched Impedance. Condition that occurs when the output impedance of a souce is equal to the input impedance of a load.

Matching. Connection of two components or circuits so that maximum power is transferred between the two.

Maximum Power Point (MPP). The point on the current-voltage (I-V) curve of a module under illumination, where the product of current and voltage is maximum. [UL 1703] For a typical silicon cell, this is at about 0.45 V.

Maximum Power Point Tracker (MPPT). Means of a power conditioning unit that automatically operates the PV-generator at its MPP under all conditions. An MPPT will typically increase power delivered to the system by 10% to 40%, depending on climate conditions and battery state of charge.

Maximum Power Transfer. A theorem that states that maximum power will be transferred from source to load when input impedance of the load equals the output impedance of the source.

Maxwell. Unit of magnetic flux. One maxwell equals one magnetic line of force.

Mbps (Mb/s). Megabits (millions of bits) per second. One megabit contains 1,048,576 bits (2^{20} bits)

MBps (MB/s). MegaBytes (millions of bytes or octets) per second. One megabyte contains 1,048,576 bytes (2^{20} bytes)

MC. *Multi-Carrier*, part of modern data transmission modulation schemes, used by GFSK, Bluetooth, etc.

MCS (broadcasting). Multipoint Communications System. An MMDS variant, sometimes called "wireless cable TV", a technology to provide television, and sometimes simplex or full duplex data services to areas where customer separation makes wired systems uneconomic. When used with a return audio channel, useful for distance learning. Operates between 2-6 GHz and above 20 GHz depending upon circumstances.

MCS. Teleconferencing use: Multipoint Conference Service.

MCU. *Multipoint Control Unit* (which see).

Medium-Earth Orbit Satellite System. A communications system with satellites in orbits about 10,000 kilometers above the Earth. Such systems include those planned by Odyssey Telecommunications International Inc., scheduled to launch in 2000, and ICO Global Communications, launching in 1998.

Megahertz (MHz). One million radio waves, or cycles, per second. Equal to one thousand Kilohertz.

Megawatt (MW). One million watts; 1000 kilowatts.

Mercury Cell. Primary cell using a mercuric oxide cathode, a zinc anode and a potassium hydroxide electrolyte.

Messaging. Synonymous with text paging, e-mail or short messages received on alphanumeric pagers and other wireless devices.

Metal Film Resistor. A resistor in which a film of metal oxide or alloy is deposited on an insulating substrate.

Metal Oxide Field Effect Transistor (MOSFET). A field effect transistor in which the insulating layer between the gate electrode and the channel is a metal oxide layer.

Metal Oxide Resistor. A metal film resistor in which an oxide of metal (such as tin) is deposited as a film onto the substrate.

Meter. Any electrical or electronic measuring device. In the metric system, it is the unit of length equal to 39.37 inches.

Meter FSD Current. Value of meter current needed to cause the needle to deflect to its maximum position (full scale deflection).

Meter Resistance. DC resistance of the meter's armature coil.

MGCP. Multimedia Gateway Control Protocol.

MHz. *MegaHertz,* in analog signals, millions of cycles or alternations per second.

Mica Capacitor. Capacitor using mica as the dialectric.

Microcell (also Picocell and Nanocell). A cell having a very small coverage area, which could be as small as one floor of an office building, one part of an airline terminal or one corner of a busy intersection. These cells are typically used where coverage and/or capacity is strained and the use of a normal sized cell would cause interference or be impractical to install. These cells transmit with extremely low power outputs.

Microcellular. A technology that directs the cellular signal into an isolated spot, leaving broader coverage to conventional cell sites.

Microgroove. A small groove scribed into the surface of a cell which is filled with metal for contacts.

Micrometer. One millionth of a meter (10-6 m).

Microphone. Electroacoustic transducer that converts sound energy into electric energy.

Microwave. Band of very short wavelength radio waves within the UHF, SHF and EHF bands.

Midband Gain. Gain of an amplifier operating within its bandwidth.

Middleware. Middleware is the 'mix-and-match' communications software that acts as a universal translator between diverse radio frequency technologies and protocols. Middleware physically re sides on the remote client and on a communications server, located between the client and the applications server. The software eases computing and communicating with corporate information and encourages applications development, making wireless data more attractive to corporate customers.

Mid-point Bias. An amplifier biased at the center of its DC load line.

Mil. One thousandth of an inch (0.001 in.)

Miller's Theorem. A theorem that allows you to represent a feedback capacitor as equivalent input and output shunt capacitors.

MIL-STD-188-131. US Military Nomenclature for H.320 compatible videoconferencing apparatus.

MIN (Mobile Identification Number). Uniquely identifies a mobile unit within a wireless carrier's network. The MIN often can be dialed from other wireless or wireline networks. The number differs from the electronic serial number, which is the unit number assigned by a phone manufacturer. MINs and ESNs can be electronically checked to help prevent fraud.

Minority Carriers. 1. The conduction band holes in n-type material and valence band electrons in p-type material. Most minority carriers are produced by temperature rather than by doping with impurities.

2. A current carrier, either an electron or a hole, that is in the minority in a specific layer of a semiconductor material; the

diffusion of minority carriers under the action of the cell junction voltage is the current in a photovoltaic device.

Minority Carrier Lifetime. The average time a minority carrier exists before recombination.

MIPS (Millions of Instructions Per Second). Used in defining digital signal processing capabilities.

Mismatch. Term used to describe a difference between the output impedance of a source and the input impedance of a load. A mismatch prevents the maximum transfer of power from source to load.

MLP. *Multi-Layer Protocol.*

MMDS. *Multi-channel Multipoint Distribution Service.* Sometimes called "wireless cable TV", a technology to provide television, and sometimes simplex or full duplex data services to areas where customer separation makes wired systems uneconomic. When used with a return audio channel, useful for distance learning.

Modem. MOdulator - DEModulator. A device that converts digital signals to a form suitable for analog transmission.

Modem Pools. Racks of modems for more reliable cellular data communications.

Modulation. Process by which an information signal (audio for example) is used to modify some characteristic of a higher frequency wave known as a carrier (radio for example).

Module. A number of PV cells connected together, sealed with an encapsulant, and having a standard size and output power; the smallest building block of the power generating part of a PV array. Also called panel.

Monolithic. Fabricated as a single structure.

Monostable Multivibrator. A multivibrator with one stable output state. When triggered, the circuit output will switch to the unstable state for a predetermined period of time and then return to the stable state. A timer.

Molecule. Smallest particle of a compound that still retains its characteristics.

Moratoria. Moratoria (the singular form is moratorium) are waiting periods on the issuance of construction permits by local zoning authorities. Moratoria are typically imposed to allow time for localities to develop or refine ordinances dealing with antenna siting issues. However, they have been used by some localities as tools to delay or block the rollout and/or expansion of wireless networks. Such usage has resulted in lawsuits by carriers.

MOSFET. Abbreviation for "metal oxide field effect transistor" also known as an "insulated gate field effect transistor). A field effect transistor in which the insulating layer between the gate electrode and the channel is a metal oxide layer.

MOS Testing. *Mean Opinion Score* Testing, a means of evaluating the subjective performance of voice and/or video transmission equipment using procedures as set out in ITU-T P.800 and related standards.

Motion Compensation. see *Motion Prediction.*

Motion Prediction. A type of interframe coding used by codecs that compress motion video. By examining statistics of previous video frame sequences, a future prediction of the motion vectors and rate-of-change for an object, block or group of pixels is developed for subsequent frames. In H.261, the estimated vectors must terminate within the image while in H.263, they may extend outside the image bounds. Important at lower transmit bit rates, optional in H.261.

MOU (Minutes of Use). A measurement of wireless subscriber activity directly affecting revenue.

Moving Coil Microphone. Microphone that uses a moving coil within a fixed magnetic field. Dynamic microphone.

Moving Coil Pick-up. Dynamic phonograph pick-up in which the stylus causes a coil to move within a fixed magnetic field.

Moving Coil Loudspeaker. Loudspeaker that uses a moving "voice coil" placed within a fixed magnetic field. Audio frequency current in the voice coil causes movement which is mechanically transferred to the speaker cone. Also known as a dynamic loudspeaker.

MPEG. Motion Pictures Experts Group, an *ISO* image standards group.

MPEG1. MPEG1 is the first of a family of motion video and audio compression standards. It provides DCT lossy compression with rather high quality. It is intended for broadcast quality applications, with resolutions based upon the CCIR (now ITU-R) 601 standard for video coding. It uses three coding protocols: Intraframe coding or I-Pictures (no prediction), P-Pictures, incorporating motion prediction from the previous video image and, B-Pictures (bi-directional prediction) which include motion prediction one frame ahead as well as from the previous frame. MPEG2 has largely supplanted MPEG1.

MPEG2. MPEG2 is the second in a family of motion video and audio compression standards. The result of a natural evolution from MPEG1, it provides DCT lossy compression ranging from low to rather high quality although not as good as MPEG1. It is intended for broadcast quality applications, with resolutions based upon the CCIR (now ITU-R) 601 standard for video coding. It uses three coding protocols: Intraframe coding or I-Pictures (no prediction), P-Pictures, incorporating motion prediction from the previous video image and, B-Pictures (bi-directional prediction) which include motion prediction one frame ahead as well as from the previous frame. MPEG2 also has a coding mode at 24 frames per second, matching that of motion picture film, useful for offline image encoding. MPEG2 has largely supplanted MPEG1 and is used for coding multimedia images from CD-ROM, DVD, broadcasting, pay TV and high quality videoconferencing. MPEG2 also includes more coding decision points than MPEG1, permitting rather fine control of image quality over time in response to the needs of other channels in applications like DBS. This standard is also known as ITU-T H.262.

MPEG3. The MPEG3 standard was intended as an ultra-high quality solution that was abandoned during development, and therefore does not exist. However, the term "MPEG3" is sometimes incorrectly used to describe MPEG2 layer 3 audio (MP3).

MPEG4. MPEG4 is an ISO standard intended to satisfy the needs of content authors, network service providers and end users. It is a multifaceted, extensible standard, and is an attempt to provide all things to all people. It covers sprites, speech synthesis, streaming video, multimedia and many other aspects including a foreground-background coding technique which will likely supplant MPEG2 for broadcast applications as it uses about one-half the bandwidth for equivalent quality. The most common version found in the PC environment is known as MPEG4 Part 3 which includes moving images and sound.

MPEG4 Part 10 AVC. An ISO motion-video compression standard using advanced video coding, a foreground-background coding technique, intended for high-quality and HDTV applications. The ISO work was merged with that of ITU-T work on same subject and is known as H.264 (which see).

MSA (Metropolitan Statistical Area). The coverage area of a city as in a cellular network. A U.S. Census Bureau term.

MSDSL. *Multi-rate Symmetrical Digital Subscriber Line,* MSDSL provides full duplex symmetrical connectivity at bit rates up to 2.0 Mbps over distances up to 8.8 Km (5.5 miles) over a single pair of telephone wires. The service provider or carrier has the capability of setting the maximum bit rate for the line, often linked to the monthly line charges. This type of transmission does not support telephony over the same physical pair as does ADSL.

MTA (Major Trading Area). MTAs are usually composed of several contiguous basic trading areas. A service area designed by Rand McNally and adopted by the FCC. There are 51 MTAs in the United States.

MTBF. Mean Time Between Failures, a measure of system or hardware reliability or availability.

MTSO (Mobile Telephone Switching Office). The electronic 'middleman' between cell sites and the public switched telephone network, processing traffic back and forth.

MTTR. Mean Time To Repair. Especially useful for in complex systems, MTTR provides a mean time from the onset of a failure to return to pre-failure operational status.

Multimeter. Electronic test equipment that can perform multiple tasks. Typically one capable of measuring voltage, current and resistance. More sophisticated modern digital multimeters also measure capacitance, inductance, current gain of transistors and/or anything else that can be measured electronically.

Multipath Propagation. Signal distortion resulting when part of a transmitted radio-frequency signal is reflected from nearby surfaces on its way to a receiver. The 'ghosting' effect on television screens illustrates the multipath phenomenon.

Multicrystalline. Material that is solidified at such as rate that many small crystals (crystallites) form. The atoms within a single crystallite are symmetrically arranged, whereas crystallites are jumbled together. These numerous grain boundaries reduce the device efficiency. A material composed of variously oriented, small individual crystals. (Sometimes referred to as polycrystalline or semicrystalline).

Multijunction Device. A photovoltaic device containing two or more cell junctions, each of which is optimized for a particular part of the solar spectrum, to achieve greater overall efficiency.

Multiplier Resistor. Resistor connected in series with a moving coil meter movement to extend the voltage ranges.

Multiplex *(verb).* To sequentially incorporate several data streams into a single data stream in such a manner that each may later be recovered intact.

Multipoint Control Unit (MCU). A device to link three or more videoconference locations for fully interactive audio, data and video teleconferencing. An ITU-T compatible MCU must meet the requirements of H.231 and H.243. In addition, the T.12x and T.13x suites of standards, desirable for both codec and MCU, provide greater functionality for multipoint videoconferencing.

Multisegmant Display. Device made of several light emitting diodes arranged in a numeric or alphanumeric pattern. By lighting selected segments numeric or alphabet characters can be displayed.

Multivibrator. A class of circuits designed to produce square waves or pulses. A stable multivibrators produce continuous pulses without an external stimulus or trigger. Monostable multivibrators produce a single pulse for some predetermind period of time only when triggered. Bistable multivibrators produce a DC output which is stable in either one of two states. Either high or low. An external stimulus or trigger is required for the bistable circuit to change states, either high to low or low to high.

Mutual Compensation. The concept that carriers must pay when they terminate traffic on the networks of carriers with which they are interconnected.

Mutually Exclusive Applications. Two or more applications for the same spectrum use rights.

Mutual Inductance. Ability of one inductor's lines of force to link with another inductor.

MVDS. Multichannel Video Distribution System. Sometimes called "wireless cable TV", a technology to provide television, and sometimes simplex or full duplex data services to areas where customer separation makes wired systems uneconomic. When used with a return audio channel, useful for distance learning. Operates between 2-6 GHz and above 20 GHz depending upon circumstances.

MWS. Multipoint Wireless System. Another name for MMDS (which see). Sometimes known as "wireless cable", a technology to provide television, and sometimes simplex or full duplex data services including Internet and VoIP to areas where customer separation makes wired systems uneconomic. When used with a return audio channel, useful for distance learning. Operates between 2-6 GHz and above 20 GHz depending upon circumstances.

N

***n*-type Semiconductor.** 1. A semiconductor compound formed by doping an intrinsic semiconductor with a pentavalent element. An n-type material contains an excess of conduction band electrons.

NAMPS (Narrowband Advanced Mobile Phone System). NAMPS combines cellular voice processing with digital signaling, increasing the capacity of AMPS systems and adding functionality.

NANC (North American Numbering Council). The FCC advisory group formerly responsible for administering the North American Numbering Plan that oversees assignment of area codes, central office codes and other numbering issues in the United States, Canada, Bermuda and part of the Caribbean. NANP administration responsibility was transferred to Lockheed Martin.

Narrowband PCS. The next generation of paging networks, including two-way, acknowledgment and 'wireless answering machine' paging.

2. A semiconductor produced by doping an intrinsic semiconductor with an electron-donor impurity (e.g., phosphorous in silicon).

NBS. The former US National Bureau of Standards. *See NIST.*

NCIC (National Crime Information Center). A national database of crime and criminal information operated by the Federal Bureau of Investigation.

NCP. Network Control Protocol (PPP usage).

NEBS. Network Equipment Building Standards. A set of standards for equipment to be installed in telco central offices.

Negative. Terminal that has an excess of electrons.

Negative Charge. A charge that has more electrons than protons

Negative Feedback. A feedback signal 180° out of phase with an amplifier input signal. Used to increase amplifier stability, bandwidth and input impedance. Also reduces distortion.

Negative Ground. A system where the negative terminal of the source is connected to the system's metal chassis.

Negative Ion. An atom having a greater number of electrons in orbit than there are protons in the nucleus.

Negative Resistance. A resistance such that when the current through it increases the voltage drop across the resistance decreases.

Negative Temperature Coefficient. A term used to describe a component whose resistance or capacitance decreases when temperature increases.

NENA (National Emergency Numbering Association). NENA's mission is to foster the technological advancement, availability and implementation of a universal emergency telephone number system.

Neon Bulb. Glass envelope filled with neon gas which when ionized by an applied voltage will glow red.

Network. Combination of interconnected components, circuits or systems.

Neutral. A terminal, point or object with balanced charges. Neither positive or negative.

Neutral Atom. An atom in which the number of negative charges (electrons in orbit) is equal to the number of positive charges (protons in the nucleus).

Neutral Wire. The conductor of a polyphase circuit or a single-phase three wire circuit that is intended to have a ground potential. The potential difference between the neutral and each of the other conductors are approximately equal in magnitude and equally spaced in phase.

Neutron. Subatomic particle in the nucleus of an atom and having no electrical charge.

NIC. Network Interface Card (PC usage, often referring to an Ethernet 10/100/1000baseT device).

Ni-Cd (Nickel Cadmium). A battery technology that is rapidly losing favor in the wireless industry. These batteries don't provide much power for wireless phones and suffer from a "memory effect" that depletes their charge very quickly if the battery is not drained all the way down before recharging.

Nickel-cadmium Cell. A secondary cell that uses a nickel oxide positive electrode and a cadmium negative electrode.

NIMBY (not in my back yard). Public sentiment that opposes local placement of 'undesirable' facilities such as antenna towers or toxic waste dumps.

NIST. US National Institute of Standards and Technology; formerly US National Bureau of Standards (NBS).

NmH (Nickel Metal Hydride). The most popular battery technology because it strikes a balance between cost and performance. Has nominal memory effect and provides significantly better standby and talk time than Ni-Cad batteries, but not quite as good as Lithium Ion batteries.

NMT (Nordic mobile telephone). An older analog cellular protocol used in Europe and elsewhere.

NOCT (Nominal Operating Cell Temperature). The solar cell temperature at a reference environment defined as 800 W/

m2 irradiance, 20□C ambient air temperature, and 1 m/s wind speed with the cell or module in an electrically open circuit state.

Node. Junction or branch point in a circuit.

Noise. Unwanted electromagnetic radiation within an electrical or mechanical system.

Noise reduction (electroacoustic). In acoustic echo cancellers (AECs), a type of filtering which reduces the amplitude of noise sources within a teleconference room transmitted to a distant conference room.

Non-inverting Input. The terminal on an operational amplifier that is identified by a plus sign.

Non-linear Scale. A scale in which the divisions are not equally spaced.

Nonrenewable Fuels. Fuels that cannot be easily made or "renewed." We can use up nonrenewable fuels. Oil, natural gas, and coal are nonrenewable fuels.

Normal Closed. Designation which states that the contacts of a switch or relay are closed or connected when at rest. When activated, the contacts open or separated.

Normally Open. Designation which states that the contacts of a switch or relay are normally open or not connected. When activated the contacts close or become connected.

North Pole. Pole of a magnet out of which magnetic lines of force are assumed to originate.

Norton's Theorem. Any network of voltage sources and resistors can be replace by a single current source in parallel with a single resistor.

Notch Filter. A filter which blocks a narrow band of frequencies and passes all frequencies above and below the band.

NOVRAM. Non-Volatile Random Access Memory. RAM that contains a constant power source. Maintains data without the presence of external power (to the end of life of the internal source.

npn Transistor. A bipolar junction transistor in which a p-type base element is sandwiched between an n-type emitter and an n-type collector.

NTIA (National Telecommunications and Information Administration). The federal government's spectrum management authority.

NTSC. The National Television System Committee 525-line color-television standard used in North America and Japan. It is synonymous with CCIR System M.

NTT. Nippon Telephone and Telegraph, the principal Japanese common carrier. As AT&T has done in the USA, so NTT has produced many of the transmission standards used in Japan.

Nuclear Energy. Energy that comes from splitting atoms of radioactive materials, such as uranium.

Nucleus. Core of an atom. The nucleus contains both positive (protons) and neutral (neutrons) subatomic particles.

Number Pooling. Increasingly popular tactic for conserving phone numbers. Numbers are returned by all carriers to a central authority, which puts them in a pool, from which carriers receive numbers in lots of 1,000, not 10,000 as was originally done. It relies on local number portability.

Numeric. A display, message or readout that contains numerals only, such as in paging.

Nyquist Criterion. From sampling theory, for a sine wave, the sampling frequency must be no less than twice the maximum signal frequency.

O

OC1. Optical Carrier level 1: 51.84 Mbps.

OC3. Optical Carrier level 3: 155.52 Mbps.

Octave. Interval between two sounds whose fundamental frequencies differ by a ratio of 2 to 1. 440 Hz. is one octave above 220 Hz.

Octet. 8-bits of data (European usage), the same as a modern byte.

Off-peak. Part of the day that wireless subscribers can expect to pay reduced airtime rates.

Offset Null. An op amp control pin used to eliminate the effects of internal component voltages on the output of the device.

Ohm. 1. Unit of resistance symbolized by the Greek capital letter omega (W). 2. The unit of resistance to the flow of an electric current.

Ohmmeter. Device used to measure electrical resistance.

Ohm's Law. Relationship between voltage, current and resistance. Ohm's law states that current in a resistance varies in direct proportion to voltage applied and inversely proportional to resistance.

Ohms Per Volt. Refers to a value of ohms per volt of full scale defection for a moving coil meter movement. The number of ohms per volt is the reciprocal of the amount of current required to produce full scale deflection of the needle. A meter requiring 50 microamps for full scale deflection has an internal resistance of 20 kW per volt. The higher the ohms per volt rating, the more sensitive the meter.

Oil. A black liquid fossil fuel found deep in the Earth. Gasoline and most plastics are made from oil.

On Circulation. A common way the FCC commissioners decide items. An item is circulated among the commissioners, allowing them to vote without having to come together as a group, which under federal law would necessitate a public meeting.

One-axis Tracking. A system capable of rotating about one axis.

1.7-way Paging. A paging service that offers more than guaranteed messaging but not as much as full two-way paging. The subscriber has limited response messaging, such as canned messages, rather than the ability to create responses.

One-shot. Monostable multivibrator.

One-stop Shop. Describes the all-in-one store where carriers sell wireless, long-distance, Internet access and any other services they are able to sell in that market.

Op-amp. Abbreviation for operational amplifier.

Open-circuit Voltage (Voc). The maximum possible voltage across a photovoltaic cell or module; the voltage across the cell in sunlight when no current is flowing.

Open Loop Gain. Gain of an amplifier when no feedback is present.

Open Loop Mode. An amplifier circuit having no means of comparing the output with the input. (No feedback.)

Operational Amplifier. A high gain DC amplifier that has a high input impedance and a low output impedance. Op-amps are the most basic type of linear integrated circuits.

Orbit. A fixed circular, elliptical or other path around the Earth.

Oscillate. To produce a continuous output waveform without an input signal present.

Oscillator. An electronic circuit that produces a continuous output waveform with only DC applied.

Oscilloscope. An instrument used to display a signal graphically. Shows signal amplitude, period and waveshape in addition to any DC voltage present. A multiple trace oscilloscope can show two or more waveforms at the same time for phase comparison and timing measurements.

OTASP (Over-the-air Service Provisioning). The ability of carriers to add new types of services to a customer's handset by using the wireless network instead of requiring the customer to bring in the phone for reprogramming.

Out of Phase. When the maximum and minimum points of two or more waveshpes do not occur at the same time.

Output. Terminal at which a component, circuit or piece of equipment delivers current, voltage or power.

Output Impedance. Impedance measured across the output terminals of a device without a load connected.

Output Power. Amount of power a component, circuit or system can deliver to a load.

Overload. Condition that occurs when the load is greater than the system was designed to handle. (Load resistance too small, load current too high.) Overload results in waveform distortion and/or overheating.

Overload Protection. Protective device such as a fuse or circuit breaker that automatically disconnects a load when current exceeds a predetermined value.

Over The Air Activation (OTAP). The process by which wireless carriers can activate a phone, assign a telephone number, change or update features and much more from a remote location by sending signals to the phone over radio frequencies. Eliminates the need for subscribers to physically take their phones into retail stores or other locations when they want to change a number or add a feature.

Overlay Area Code. A solution to the scarcity of new phone numbers, overlays involve issuance of new 10-digit phone numbers for use alongside an area's existing seven-digit numbers, which have a different area code.

P

P.800. Formerly known as P.80, an ITU-T recommendation regarding methods for the subjective determination of transmission quality and forms the basis of a wide range of subjective quality and intelligibility testing. These tests are offered by HEI.

P.830. This ITU-T recommendation covers subjective performance assessment of telephone-bandwidth and wideband digital codecs.

P.861. The objective quality measurement of telephone bandwidth (300Hz - 3400Hz) speech codecs is the topic of this ITU-T recommendation. The tests need to be applied with care as they produce valid results with only a limited range of codecs and test conditions. See PSQM below.

P.910. An ITU-T recommendation on the subjective assessment of video transmission quality for multimedia applications. Often, the criteria to be evaluated are determined using RISV or VIRIS.

P.920. An ITU-T recommendation regarding interactive test methods for the quality assessment of audio and visual communications, tests offered by HEI.

P.930. An ITU-T recommendation setting out the principles of a reference impairment system for use with subjective testing methods of the quality of various compressed video signals.

P802.9a. An IEEE standard providing the overview and architecture of: Integrated Services (IS) LAN: IEEE 802.9 Isochronous services with Carrier sense multiple access with collision detection (CSMA/CD) Media Access Control (MAC) service, the Iso-Ethernet standard. See also H.322.

Packet. A form of data transmission dividing the data information into many small packets, each including information such as source, destination, protocol and packet length. The concept is used for the Internet where transmission facilities are shared by many different users, with packets removed or added as appropriate at different locations.

Packing Factor. The ratio of array area to actual land area or building envelope area, for a system; or, the ratio of total PV cell area to the total module area, for a module.

PACS (Personal Access Communications System). An extended personal cordless technology developed by Hughes Network Systems Inc. and Bellcore, planned for implementation by C-Block licensee 21st Century Telesis.

PAD. Packet Assembler/Disassembler.

Pager. A small device capable of receiving numbers and, in some cases, text messages that are sent through a telephone connection or a computer. The earliest pagers were only one-way, i.e., they could only receive information. More recent models allow the pager user to reply back to the sender, called 2-way paging. Some pagers cannot respond, but have the benefit of guaranteeing that the page will be received because they send a brief acknowledgement back to the system when they receive a page. Until that acknowledgement is sent, the system will continue sending the page repeatedly.

PAL. Phase Alternate Line, the European 625-line, 25-frame per second color-television standard.

PAM. Pulse Amplitude Modulation, a sampled analog signal which may exist as an intermediate step prior to the signal's coding and quantizing.

PAMR (Public Access Mobile Radio). The European designation for services similar to specialized mobile radio in the United States.

Panel. See 'Photovoltaic panel'.

Paper Capacitor. Fixed capacitor using oiled or waxed paper as a dialectric.

Parallel. Circuit having two or more paths for current flow. Also called shunt.

Parallel Connection. A way of joining two or more electricity-producing devices (i.e., PV cells or modules) by connecting positive leads together and negative leads together; such a configuration increases the current.

Parallel Resonant Circuit. Circuit having an inductor and a capacitor in parallel with one another. Circuit offers a high impedance at resonant frequency. Sometimes called a "tank circuit."

Partitioning. Parceling a spectrum license into two or more geographic areas.

Pass Band. The range of frequencies that will be passed and amplified by a tuned amplifier. Also the range of frequencies passed by a band pass filter.

Passivation. A chemical reaction that eliminates the detrimental effect of electrically reactive atoms on a photovoltaic cell's surface.

Passive Component. Component that does not amplify a signal. Resistors and capacitors are examples.

Passive Filter. A filter that contains only passive or non amplifying components.

Passive Solar Home. A house that uses a room or another part of the building as a solar collector.

Passive System. System that emits no energy. It only receives. It does not transmit or reveal its position.

Pay Phone Service Provider. The company that owns and operates a pay phone.

PBX. *Private Branch eXchange.* A telephony/data switching system, usually located on customer premises and belonging to the user. In contrast to Centrex which is largely located on the common carrier's premises. Some modern PBXs provide effective switched digital interfaces for operation of switched 64/56 Kbps and ISDN videoconferencing.

Some contemporary PBXs also provide VoIP encoding for VPN communications between an organization's offices.

PCB. *Printed Circuit Board.* An electronic assembly which is plugged into the main frame assembly of a complex electronic device or which may be used as the principle element of a simple electronic device.

PC Card. Formerly called a PCMCIA card, a detachable card that can be connected to the mother board inside a personal computer. It is used to link the PC to other devices to carry out a special function.

PCIA (Personal Communications Industry Association). A trade group representing PCS, SMR, private radio and other wireless users and carriers.

PCM. *Pulse Code Modulation,* a basic form of digital modulation where an analog signal is sampled with each sample being quantized independently of the other samples.

PCS (Personal Communications Services). A two-way, 1900 MHz digital voice, messaging and data service designed as the second generation of cellular.

PDA (Personal Digital Assistant). A portable computing device capable of transmitting data.

PDU. Protocol Data Unit.

Peak. 1. Maximum or highest amplitude level. 2. Part of the day that mobile phone customers can expect to pay full service airtime rates.

Peak Inverse Voltage (PIV). The maximum rated value of a AC voltage acting in the direction opposite to that in which a device is designed to pass current.

Peak Load; Peak Demand. The maximum load, or usage, of electrical power occurring in a given period of time, typically a day.

Peak Power. Power generated by a utility unit that operates at a very low capacity factor; generally used to meet short-lived and variable high demand periods.

Peak to Peak. Difference between the maximum positive and maximum negative values of an AC waveform.

Peak Watts (Wp). See 'Photovoltaic peak watt'.

PEL. *Picture ELement.*

Pentavalent Element. Element whose atoms have five valence electrons. Used in doping intrinsic silicon or germanium to produce n-type semiconductor material. Most commonly used pentavalent materials are arsenic and phosphorus.

Percent of Regulation. The change in output voltage that occurs between no-load and full-load in a DC voltage source. Dividing this change by the full-load value and multiplying the result by 100 gives percent regulation.

Percent of Ripple. The ratio of the effective rms value of ripple voltage to the average value of the total voltage. Expressed as a percentage.

Period. Time to complete one full cycle of a periodic or repeating waveform.

Permanence. Magnetic equivalent of magnetic inductance and consequently equal to the reciprocal of reluctance, just as conductance is equal to the reciprocal of resistance.

Permanent Magnet. Magnet normally made of hardened steel that retains its magnetism indefinately.

Permeability. Measure of how m how much better a material is as a path for magnetic lines of force with respect to air which has a permeability of one. Symbolized by the Greek lower case letter mu (m).

Phase. Angular relationship between two waves.

Phase Angle. Phase difference between two or more waves, normally expressed in degrees.

Phase Shift. Change in phase of a wave form between two points, expressed as degrees of lead or lag.

Phase Shift Oscillator. An oscillator that uses three RC networks in its feedback path to produce the 180□ phase shift required for oscillation.

Phase Splitter. Circuit that takes a single input signal and produces two output signals that are 180□ apart in phase.

Phonograph. Piece of equipment used to reproduce sound stored on a disk called a phonograph record.

Phosphor. Luminescent material applied to the inner face of a cathode ray tube that when bombarded with electrons will emit light of various colors.

Phosphorous (P). A chemical element, atomic number 15, used as a dopant in making n-semiconductor layers.

Photoconductive Cell. Material whose resistance decreases or conductance increases when exposed to light.

Photoconduction. A process by which the conductance of a material is change by incident electromagnetic radiation in the visible light spectrum.

Photocurrent. An electric current induced by radiant energy.

Photodetector. Component used to detect or sense light.

Photodiode. A semiconductor diode that changes its electrical characteristics in response to illumination.

Photoelectrochemical Cell. A special kind of photovoltaic cell in which the electricity produced is used immediately within the cell to produce a useful chemical product, such as hydrogen. The product material is continuously withdrawn from the cell for direct use as a fuel or as an ingredient in making other chemicals, or it may be stored and used subsequently.

Photometer. Meter used to measure light intensity.

Photon. 1. Discrete portion of electromagnetic energy. A small packet of light. 2. A particle of light that acts as an individual unit of energy.

Photoresistor. Also known as a photoconductive cell or light dependent resistor. (LDR) A device whose resistance decreases with exposure to light.

Photovoltaic (PV). Pertaining to the direct conversion of light into electricity.

Photovoltaic (PV) Array. An interconnected system of PV modules that function as a single electricity-producing unit. The modules are assembled as a discrete structure, with common support or mounting. In smaller systems, an array can consist of a single module.

Photovoltaic Cell. 1. Component commonly called a solar cell used to convert light energy into electrical energy. 2. The smallest semiconductor element within a PV module to perform the immediate conversion of light into electrical energy (dc voltage and current).

Photovoltaic (PV) Conversion Efficiency. The ratio of the electric power produced by a photovoltaic device to the power of the sunlight incident on the device.

Photovoltaic (PV) Device. A device that converts light directly into dc electricity.

Photovoltaic (PV) Efficiency. The ratio of electric power produced by a cell at any instant to the power of the sunlight striking the cell. This is typically about 9% to 14% for commercially available cells.

Photovoltaic (PV) Generator. The total of all PV strings of a PV power supply system, which are electrically interconnected.

Photovoltaic (PV) Module. The smallest environmentally protected, essentially planar assembly of solar cells and ancillary parts, such as interconnections, terminals, [and protective devices such as diodes] intended to generate dc power under unconcentrated sunlight. The structural (load carrying) member of a module can either be the top layer (superstrate) or the back layer (substrate). [UL 1703]

Photovoltaic (PV) Panel. often used interchangeably with PV module (especially in one-module systems), but more accurately used to refer to a physically connected collection of modules (i.e., a laminate string of modules used to achieve a required voltage and current).

Photovoltaic (PV) Peak Watt. Maximum "rated" output of a cell, module, or system. Typical rating conditions are 0.645 watts per square inch (1000 watts per square meter) of sunlight, 68 degrees F (20 degrees C) ambient air temperature and 6.2 x 10-3 mi/s (1 m/s) wind speed.

Photovoltaic (PV) System. A complete set of components for converting sunlight into electricity by the photovoltaic process, including the array and balance of system components.

Photovoltaic-thermal (PV/T) System. A photovoltaic system that, in addition to converting sunlight into electricity, collects the residual heat energy and delivers both heat and electricity in usable form. Also called a total energy system.

PHS (Personal Handyphone System). The extended cordless system used primarily in Japan.

Physical Vapor Deposition. A method of depositing thin semiconductor films. With this method, physical processes, such as thermal evaporation or bombardment of ions, are used to deposit elemental semiconductor material on a substrate.

Pi. Value representing the ratio between the circumference and diameter of a circle and equal to approximately 3.142.

Picture Element. The smallest discrete part of a video image, the size of which is controlled by the analog-to-digital conversion sampling process and subsequent other compression processes. The more picture elements per line, the higher the resolution of the image. To convert the number of H pixels to a close approximation of TV lines of resolution for the NTSC system, simply multiply the number of H pixels by 0.78. Furthermore, each 80 TV lines of resolution (NTSC) requires 1 MHz of analog bandwidth for transmission.

PID. *Premises Interface Device,* a filtering and splitter device for interfacing bidirectional cable systems and the customer.

Pierce Oscillator. A variation of the colpitts oscillator. This oscillator uses a quartz crystal in place of the inductor found in the colpitts oscillator feedback network. The crystal maintains a highly stable output frequency.

Piezoelectric Crystal. Crystal material that will generate a voltage when mechanical pressure is applied and conversely will undergo mechanical stress when subjected to a voltage.

Piezoelectric Effect. The production of a voltage between opposite sides of a piezoelectric crystal as a result of pressure or twisting. Also the reverse effect which the application of a voltage to opposite sides causes a deformation to occur at the frequency of the applied voltage. (Converts mechanical energy into electrical energy and electrical energy into mechanical energy.)

p-i-n. A semiconductor device structure that layers an intrinsic semiconductor between a p-type semiconductor and an n-type semiconductor; this structure is most often used with amorphous silicon devices.

PIN (Personal Identification Number). A code used by a mobile telephone number in conjunction with an SIM card to complete a call.

Pinch-off Region. A region on the characteristic curve of a FET in which the gate bias causes the depletion region to extend completely across the channel.

PIP. Picture-In-Picture display. A video display mode in which a one-quarter- or smaller-size video image is superimposed over a full-screen video image. PIP is especially useful as a preview monitor for one-monitor videoconferencing installations and for TV receivers.

Pixel. *Picture element*.

Plastic Film Capacitor. Capacitor in which alternate layers of aluminum foil are separated by thin films of plastic dialectric.

Plate. Conductive electrode in either a capacitor or battery. In vacuum tube technology, it is the name given to the anode.

Plug. Movable connector that is normally connected into a socket or jack.

p/n. A semiconductor device structure in which the junction is formed between a p-type layer and an n-type layer.

pnp Transistor. A bipolar junction transistor with an n-type base and p-type emitter and collector.

POCSAG. Referring to a standard developed by the U.K.'s Post Office Code Standards Advisory Group, a paging protocol.

Point-contact Cell. A high efficiency silicon concentrator cell that employs light trapping techniques and point-diffused contacts on the rear surface for current collection.

Pole. In an active filter, a single RC circuit. A one pole filter has one capacitor and one resistor. A two pole filter has two RC circuits and so on.

Polar Coordinates. Either of two numbers that locate a point in a plane by its distance from a fixed point and the angle this line makes with a fixed line.

Polarity. Term used to describe positive and negative charges.

Polarized. A component which must be connected in correct polarity to function and/or d/or prevent destruction. Example: Electrolytic capacitor.

Polycrystalline. See 'Multicrystalline'.

Pool Consolidation. The restructuring of 20 private land mobile services into two pools—public safety and industrial/ business—during the commission's ongoing refarming proceeding.

POP. *Point-Of-Presence,* the common carrier physical location in a city, for example.

POPs (Persons of Population). Wireless industry term for the number of potential subscribers within the licensed area of a cellular or PCS system.

Positive. Polarity of point that attracts electrons as opposed to negative which supplies electrons.

Positive Charge. A charge that exists in a body that has fewer electrons than protons.

Positive Feedback. A feedback signal that is in phase with an amplifier input signal. Positive feedback is necessary for oscillation to occur.

Positive Ground. A system whereby the positive terminal of the source is connected to the system's conducting chassis.

Positive Ion. Atom that has lost one or more valence electrons resulting in a net positive charge.

Potential Difference. Voltage difference between two points which will cause current to flow in a closed circuit.

Potential Energy. Energy that has potential to do work because of its position relative to others.

Potentiometer. A variable resistor with three terminals. Mechanical turning of a shaft can be used to produce variable resistance and potential. Example: A volume control is usually a potentiometer.

POTS. Jargon, *Plain Old Telephone Service.*

Power. Amount of energy converted by a circuit or component in a unit of time, normally seconds. Measured in units of watts. (joules/second).

Power Amplifier. An amplifier designed to deliver maximum power output to a load. Example: In an audio system, it is the power amplifier that drives the loudspeaker.

Power Conditioning Equipment. Electrical equipment, or power electronics, used to convert power from a photovoltaic array into a form suitable for subsequent use. A collective term for inverter, converter, battery charge regulator, and blocking diode.

Power Density. The ratio of the power available from a battery to its mass (W/kg) or volume (W/l).

Power Derating Factor. A transistor rating that tells how much the maximum allowable value of P_D decreased for each 1□C rise in ambient temperature.

Power Dissipation. Amount of heat energy generated by a device in one second when current flows through it.

Power Factor. 1. Ratio of actual power to apparent power. 2. The ratio of the average power and the apparent volt-amperes.

Power Loss. Ratio of power absorbed to power delivered.

Power Supply. Electrical equipment used to deliver either AC or DC voltage.

Power Supply Rejection Ratio. A measure of an op-amps ability to maintain a constant output when the supply voltage varies.

PPP. Point-to-Point Protocol, often used for Internet access.

PPTP. Point-to-Point Tunneling Protocol, a newer technology supporting multi-protocol virtual private networks (VPNs), letting remote user access to corporate networks securely across the Internet.

PRA. European name for ISDN PRI.

Pre-emption. A federal agency voiding a local ordinance or state law, asserting that the federal government, not the state or locality, has ultimate jurisdiction on the matter.

Prepaid Cellular. A system allowing subscribers to pay in advanced for wireless service. Prepaid is generally used for credit-impaired customers or those who want to adhere to a budget.

PRI. North American Primary-Rate Interface, 23B+D ISDN interface, made up of a 64 Kbps D-channel for signalling, and 23-B channels for data, each operating at 64 Kbps.

Price/earnings Ratio. A measurement of a company's value as expressed by the stock market. P/E is a company's stock price divided by its net income per share. When a company has a high P/E, that typically means investors are willing to pay a premium for its stock in anticipation that net income will continue to grow at a certain pace. High P/Es are emblematic of established growth companies such as Microsoft. Companies with lower P/Es would include banks, the Bell phone companies and other sectors with slower growth rates. Low P/E also could signal a company is viewed negatively by investors.

Primary. First winding of a transformer. Winding that is connected to the source as opposed to secondary which is a winding connected to a load.

Primary Cell. Cell that produces electrical energy through an internal electrochemical action. Once discharged a primary cell cannot be reused.

Printed Circuit Board. Insulating board containing conductive tracks for circuit connections.

Programmable UJT. Unijunction transistor with a variable intrinsic stand-off ratio.

Project 25. An APCO-sponsored project to ensure interoperability of 800 MHz trunked Public Safety communications systems produced by different manufacturers.

Project Angel. AT&T Corp.'s code name for a wireless local loop residential network.

Projected Area. The net south-facing glazing area projected on a vertical plane.

Propagation. Travelling of electromagnetic, electrical or sound waves through a medium.

Propagation Delay. Time required for a signal to pass through a device or circuit.

Propagation Time. Time required for a wave to travel between two points.

Proportional and Owned Subscribers. Since many wireless carriers share certain markets in partnership with other carriers, they report subscribership for those markets in proportion to their share of the venture (e.g., their proportionate share of 10,000 subscribers in a market they share 50-50 with a partner would be 5,000). Conversely, in markets they operate solely, carriers report all of their subscribers as their own.

Protection Channel. A standby, fully operational facility, providing diversity for one or several similar channels. When correctly implemented, it can result in improved availability or reliability of the channel, and of the circuits carried on each so protected channel.

Protoboard. Board with provision for attaching components without solder. Also called a breadboard. Primarily used for constructing experimental circuits.

Proton. Sub atomic particle within the nucleus of an atom. Has a positive charge.

PSAP (Public-safety Answering Point). The dispatch office that receives 911 calls from the public. A PSAP may be local fire or police department, an ambulance service or a regional office covering all services.

PSDN. Packet Switched Digital Network.

PSP. A company that owns or operates pay phones.

PSQM. *Perceptual Speech Quality Measurement.* Using artificial speech, this test provides numeric values of *approximate* speech intelligibility including effects such as noise, coding errors, packet reordering, phase jitter, and excessive bit error rate. The value PSQM=0 signifies no impairment, while PSQM=6.5 is totally unusable. HEI offers these tests. See also P.861

PSTN (Public Switched Telephone Network). The worldwide voice telephone system, also called the Bell System in the United States.

PSWAC (Public Safety Wireless Advisory Committee). The FCC group that identified the safety community's wireless needs, motivating the commission's decision to reallocate 24 megahertz currently used by broadcasters to public safety agencies.

PTT. Posts, Telephone and Telegraph administration, international name for telephone companies.

P-type Semiconductor. A semiconductor in which holes carry the current; produced by doping an intrinsic semiconductor with an electron acceptor impurity (e.g., boron in silicon).

PUC (Public Utility Commission). The general name for the state regulatory body charged with regulating utilities including telecommunications.

Pulse. Rise and fall of some quantity (usually voltage) for a period of time.

Pulse Fall Time. Time for a pulse to decrease from 90% of its peak value to 10% of its peak value.

Pulse Repetition Frequency. The number of times per second that a pulse is transmitted. Pulse rate.

Pulse Repetition Time. Time interval between the start of two consecutive pulses.

Pulse Rise Time. Time required for a pulse to increase from 10% of its peak value to 90% of its peak value.

Pulse Width. Time interval between the leading edge and trailing edge of a pulse at a point where the amplitude is 50% of the peak value.

Pulse-width-modulated (pwm) Wave Inverter (PWM). PWM inverters are the most expensive, but produce a high quality of output signal at minimum current harmonics. The output voltage is very close to sinusoidal.

Punch List. The list of sophisticated wiretapping function that the FBI wants common carriers to provide under the 1994 digital wiretap law, but which the carriers say is too costly and may exceed the law's scope. The FCC has been asked to decide whether the industry's standard is sufficient.

Push>push-pull Amplifier. Amplifier using two active devices operating 180□ apart.

Push-to-talk (PTT). Works with half-duplex devices to transmit and receive voice traffic over the same frequency. CB radios, police radios and walkie-talkies designed for family use would all be examples of PTT.

PV. Abbreviation for photovoltaic(s).

PVC. Permanent Virtual Circuit (frame relay use).

PWT. *Personal Wireless Technology*, the US version of DECT, operating between 1910 and 1920 MHz.

Pyronometer. An instrument for measuring total hemispherical solar irradiance on a flat surface, or "global" irradiance; thermopile sensors have been generally identified as pyranometers, however, silicon sensors are also referred to as pyranometers.

Pythagorean Theorem. A theorem in geometry: The square of the hypotenuse of a right triangle equals the sum of the squares of the other two sides. In electronics used for vector analysis of AC circuits.

Q

Q. Quality factor of an inductor or capacitor. It is the ratio of a component's reactance (energy stored) to its effective series resistance (energy dissipated). For a tuned circuit, a figure of merit used in bandwidth calculations. Q is the ratio of reactive power to resistive power in a tuned circuit. Also the symbol for charge in coulombs (Q for quantity).

QAM. *Quadrature Amplitude Modulation.*

QCIF. *Quarter Common Intermediate Format.* Describes a subset of the type of coded video signals transmitted when using ITU-T H.261 and H.263 coding methods. See *CIF.*

QoS. *Quality of Service* which includes technical performance parameters like differential phase, latency, phase jitter or error distribution, together with worst-case performance values for each.

QPSK. *Quadrature Phase Shift Keying,* a modulation method used for digital satellite transmission.

Quad. A measure of energy equal to one trillion Btus; an energy equivalent to approximately 172 million barrels of oil.

Qualification Test (PV). A procedure applied to a selected set of PV modules involving the application of defined electrical, mechanical, or thermal stress in a prescribed manner and amount. Test results are subject to a list of defined requirements.

Quantization. The process of dividing a continuous range of signal values into larger, less granular, contiguous parts, a unique value being assigned to each part. The process assigns the same digital value to two or more of the original adjacent digital values. Quantizing is used as part of a lossy coding algorithm.

Quantizers may be linear, non-linear or dynamic, depending upon the circumstances and application.

- Y

Red minus luminance (Y). One of the two color signals used in encoding television signals.

Quiescent. At rest. For an amplifier the term is used to describe a condition with no active input signal.

Quiescent Point (Q point). A point on the DC load line of a given amplifier that represents the quiescent (no signal) value of output voltage and current for the circuit.

R

Radar. Acronym for "radio detection and ranging" A system that measures the distance and direction of objects.

Radioactive Waste. Materials left over from making nuclear energy. Radioactive waste can harm people and the environment if it is not stored safely.

Radioastronomy. Branch of astronomy that studies the radio waves generated by celestial bodies and uses these emissions to obtain information about them.

Radio Broadcast. Transmission of music, voice and other information on radio carrier waves that can be received by the general public.

Radio Common Carrier (RCC). An independent company licensed by the FCC to provide a service to the general public that uses the radio spectrum. In the early 1980s, RCCs were also referred to as "non-wireline' companies and were afforded one of two licenses for the original cellular spectrum in each market in the United States.

Radiocommunication. Term used to describe the transfer of information between two or more points by use of radio or electromagnetic waves.

Radio-frequency Amplifier. Amplifier having one or more active devices to amplify radio signals.

Radio-frequency Fingerprinting. A process that identifies a cellular phone by the unique 'fingerprint' that characterizes its signal transmission. RF fingerprinting is one process used to prevent cloning fraud, since a cloned phone will not have the same fingerprint as the legal phone with the same electronic identification numbers.

Radio-frequency Generator. Generator capable of supplying RF energy at any desired frequency in the radio-frequency spectrum.

Radio-frequency Probe. Probe used in conjunction with an AC meter to measure radio-frequency signals.

RADSL. *Rate-Adaptive Digital Subscriber Line*, a form of xDSL, it provides full duplex asymmetrical connectivity at 7.0 Mbps from the network, and 1 Mbps to the network over distances up to 5.5 Km (3.4 miles). This type of transmission supports telephony over the same physical pair as does ADSL. Like a standard dial-up modem, RADSL can change its maximum bit rate in response to changing line conditions.

Rapid Deployment System. A telecommunications system often including videoconferencing and a variety of data capabilities, packaged in air-transportable containers and permitting rapid setup. These systems are intended for establishing telecommunications and videoconferencing in emergency and crisis situations.

Rate Center. The geographic area used by local exchange carriers to set rate boundaries for billing and for issuing phone numbers. Wireless industry groups decry the rate center concept as wasteful of phone numbers because the concept is issued over larger areas.

RC. Abbreviation for "resistance capacitance" also abbreviation for "radio controlled" as in "RC model airplanes."

RBOC (Regional Bell Operating Company). The list of such companies includes Bell Atlantic, U S West, Ameritech, Southwestern Bell and BellSouth.

RC Time Constant. Product of resistance and capacitance in seconds.

Reactance. Symbol "X". Opposition to current flow without the dissipation of energy. Example: The opposition provided by inductance or capacitance to AC current.

Reactive Power. Also called imaginary power or wattless power. It is the power value in "volt amps" obtained from the product of source voltage and source current in a reactive circuit.

Real Number. Number having no imaginary part.

Receive-Only Codec. A video codec able only to receive communications signals and process them for local output. Primarily used at receive locations in point-to-multipoint or broadcast applications such as *DBS* where two-way communication with the sending location is not required. Security systems also use receive-only codecs.

Receiver. Unit or piece of equipment used to receive information.

Reciprocal Billing. The 1996 Telecommunications Act mandated that wireline companies pay wireless companies for their cost of terminating calls that originated on the wireline network. Previously, only wireless companies were obligated to pay compensation for calls originated on their networks but terminated on the wireline network.

Recombination. 1. Process by which a conduction band electron gives up energy (in the form of heat or light) and falls into a valence band hole.

2. The action of a free electron falling back into a hole. Recombination processes are either radiative, where the energy of recombination results in the emission of a photon, or nonradiative, where the energy of recombination is given to a second electron which then relaxes back to its original energy by emitting phonons. Recombination can take place

in the bulk of the semiconductor, at the surfaces, in the junction region, at defects, or between interfaces.

Recon Petition. Petition for reconsideration of an FCC decision.

Rectangular Coordinates. A Cartesian coordinate of a Cartesion coordinate system whose straight-line axes or coordinate planes are perpendicular.

Rectangular Wave. Also known as a pulse wave. A repeating wave that only operates between two levels or values and remains at one of these values for a small amount of time relative to the other value.

Rectification. Process that converts alternating current to direct current.

Rectifier. 1. Diode circuit that converts alternating current into pulsating direct current. 2. A device that converts ac to dc. See 'inverter'.

Recursive filtering. A method of motion video noise reduction developed by the BBC; the hardware was known as a Digital Noise Reducer (DNR), manufactured by Pye (UK). Random noise is identified by its lack of connection to similar values of picture elements adjacent to it or, between successive video frames and at similar coordinates. Such pixels identified as noise may then be removed by replacing them with the average value of the surrounding pixels.

Recycling. A way to reuse materials instead of just throwing them away.

Recycle Mark. A design of three arrows that make up a circle. This mark tells you that you can recycle the product. It can also mean that the material is made from recycled materials.

Redundancy (telecommunications). Alternative and/or duplicate transmission paths, routes, equipment and power in various combinations to enhance the reliability of a telecommunications infrastructure.

Redundant (information). Video data information that does not change over a certain time interval (temporally redundant) or video data information where a given pixel is surrounded with similar pixels (spatially redundant).

Reed Relay. Relay consisting of two thin magnetic strips within a glass envelope. When a coil around the envelope is energised, the relays contacts snap together making a connection between leads attached to the reed strips.

Reed-Solomon Code. A forward-error correction technique used for satellite transmission and CD recordings, especially effective for error bursts.

Refarming. An FCC initiative to promote more efficient use of the frequency bands below 512 MHz, allocated to private land mobile radio services.

ReFLEX 25. The narrowband PCS technology developed by Motorola that allows for two-way text messaging.

ReFLEX 50. The technology that denotes the proprietary two-way system implemented by SkyTel Communications Inc.

Regenerative Feedback. Positive feedback. Feedback from the output of an amplifier to the input such that the feedback signal is in phase with the input signal. Used to produce oscillation.

Regulated Power Supply. Power supply that maintains a constant output voltage under changing load conditions.

Regulator. Device or circuit that maintains a desired output under changing conditions.

Relay. Electromechanical device that opens or closes contacts when a current is passed through a coil.

Relative. Not independent. Compared with or with respect to some other measured quantity.

Relaxation Oscillator. Free running circuit that outputs pulses with a period dependent or one or more RC time constants.

Reluctance. Resistance to the flow of magnetic lines of force.

Remanence. Amount a material remains magnetized after the magnetizing force has been removed.

Remote Systems. Systems off of the utility grid.

Renewable Fuels. Fuels that can be easily made or "renewed." We can never use up renewable fuels. Types of renewable fuels are solar, wind, and hydropower energy.

Repeater. Devices that receive a radio signal, amplify it and re transmit it in a new direction. Used in wireless networks to extend the range of base station signals, thereby expanding coverage-within limits-more economically than by building additional base stations. Repeaters typically are used for buildings, tunnels or difficult terrain.

RES. Acronym for the Residential Experiment Stations.

Reseller. An independent company that buys airtime in bulk at wholesale rates from carriers and then sells the airtime directly to individuals or companies. The difference between the wholesale and retail rate is their margin and they keep whatever portion of the margin they can after paying for all other operating expenses, such as billing, customer service, etc.

Reserve Capacity. The amount of generating capacity a central power system must maintain to meet peak loads.

Residual Magnetism. Magnetism remaining in the core of an electromagnet after the coil current is removed.

Residuals. Funds retailers receive after a cellular customer has activated service; the funds usually are calculated based on the customer's ongoing service.

Resistance. Symbolized "R" and measured in ohms. Opposition to current flow and dissipation of energy in the form of heat.

Resistive Power. Amount of power dissipated as heat in a circuit containing resistive and reactive components. True power as opposed to reactive power.

Resistive Temperature Detector (RTD). Temperature detector consisting of a fine coil of conducting wire (such as platinum) that will produce a relatively linear increase in resistance as temperature increases.

Resistive Voltage Drop. The voltage developed across a cell by the current flow through the resistance of the cell.

Resistivity. Measure of a material's resistance to current flow.

Resistor. Component made of material that opposes flow of current and therefore has some value of resistance.

Resistor Color Code. Coding system of colored stripes on a resistor to indicate the resistor's value and tolerance.

Resonance. Circuit condition that occurs at the frequency where inductive reactance (X_L) equals capacitive reactance (X_C).

Restricted Channel. A digital communications channel for which each p or B channel (ITU-T usage) gives a usable capacity of 56 Kbps instead of 64 Kbps. Still sometimes used in North America, this is a remnant of the ones-density limitation of early T1 carrier facilities.

Reverse Bias. 1. Bias on a PN junction that allows only leakage current (minority carriers) to flow. Positive polarity on the n-type material and negative polarity to the p-type material.

2. Condition where the current producing capability of a PV cell is significantly less than that of other cells in its series string. This can occur when a cell is shaded, cracked, or otherwise degraded or when it is electrically poorly matched with other cells in its string.

Reverse Breakdown Voltage. Amount of reverse bias that will cause a PN junction to break down and conduct in the reverse direction.

Reverse Current. Current through a diode when reverse biased. An extremely small current also referred to as leakage.

Reverse Saturation Current. Reverse current through a diode caused by thermal activity. This current is not affected by the amount of reverse bias on the component, but does vary with temperature.

RF. Abbreviation for "radio frequency."

RFI (project management usage). Request For Information.

RFI (electro-magnetic usage). Radio Frequency Interference.

RFP. Request For Proposal.

RFQ. Request For Quotation.

RGB Video Signal. A video signal where the red (R), green (G) and blue (B) picture components are present as individual signals. Synchronization information may be separate or included with the G signal.

RGB/N. An RGB video signal based on the NTSC 525-line scanning rates.

RGB/P. An RGB video signal based on the PAL 625-line scanning rates.

Rheostat. Two terminal variable resistor used to control current.

Ribbon. A thin sheet of crystalline or multicrystalline material, such as silicon, produced in a continuous process by withdrawal from a molten bath of the parent material.

Right Angle Triangle. Triangle having a 90□ or square corner.

Ripple Frequency. Frequency of the ripple present in the output of a DC source.

Ripple Voltage. The small variations in Dc voltage that remain after filtering in a power supply.

Rise Time. Time for the leading edge of a pulse to rise from 10% of its peak value to 90% of its peak value.

RISV. Reference Impairment System for Video. RISV is used to simulate video compression impairments, independent of the compression scheme in preparation for P.910 video subjective testing.

RJ-11. USOC Registered Jack #11; a North American standard 4-wire modular telephone jack.

RL Differentiator. An RL circuit whose output voltage is proportional to the rate of change of the input voltage.

RL Filter. Selective circuit of resistors and inductors that offers little or no opposition to certain frequencies while blocking or attenuating other frequencies.

RL Integrator. RL circuit with an output proportionate to the integral of the input signal.

rms. Abbreviation for "root mean square"

rms Value. Rms value of an AC sine wave is 0.707 times the peak value. This is the effective value of an AC sine wave. The rms value of a sine wave is the value of a DC voltage that would produce the same amount of heat in a heating element.

Roaming. Traveling outside a carrier's local area.

Roll-off Rate. Rate of change in gain when an amplifier is operated outside of its bandwidth.

ROM. Read-Only Memory. A type of semiconductor memory device that stores unalterable data or program information. ROMs are best suited to large volume applications, and are often programmed using a custom mask as part of the IC lithography process.

Rotary Switch. Electromechanical device that has a rotating shaft connected to one terminal capable of making or breaking a connection to one or more other terminals.

R-2R Ladder. Network or circuit composed of a sequence of L networks connected in tandem. Circuit used in digital to analog converters.

RSA (Rural Service Area). Designation of a non-metropolitan area covered by a cellular licensee.

RTP. Real-time Transport Protocol.

RTSP. Real-Time Streaming Protocol.

S

Satellite Phone. A wireless phone that uses mobile satellite service to send voice and data.

Satellite Power System (SPS). Concept for providing large amounts of electricity for use on the Earth from one or more satellites in geosynchronous Earth orbit. A very large array of solar cells on each satellite would provide electricity, which would be converted to microwave energy and beamed to a receiving antenna on the ground. There, it would be reconverted into electricity and distributed the same as any other centrally generated power, through a grid.

Saturation. Condition in which a further increase in one variable produces no further increase in the resultant effect. In a bipolar junction transistor, the condition when the emitter to collector voltage is less than the emitter to base voltage. This condition puts forward bias on the base to collector junction.

Sawtooth Wave. Repeating waveform that rises from zero to maximum value linearly drops back to zero and repeats. A ramp waveform.

S-Band. The frequency spectrum near 2 GHz used for land based microwave and some mobile satellite communications.

Scale. Set of markings used for measurement.

Schematic Diagram. Illustration of an electrical or electronic circuit with the components represented by their symbols.

Schmitt Trigger. Circuit to convert a given waveform to a square wave output.

Schottky Barrier. A cell barrier established as the interface between a semiconductor, such as silicon, and a sheet of metal.

Schottky Diode. High speed diode that has very little junction capacitance. Also known as a "hot-carrier diode" or a "surface-barrier diode."

Scientific Notation. Numbers entered as a number from one to ten multiplied by a power of ten. Example: 8765 = 8.765 □ 10^3.

Scribing. The cutting of a grid pattern of grooves in a semiconductor material, generally for the purpose of making interconnections.

SDSL. *Symmetrical Digital Subscriber Line*, symmetrical, like HDSL but operating over a single pair, SDSL provides full duplex symmetrical connectivity at 2.3 Mbps over distances up to 6.7 Km (4.2 miles). This type of transmission does not support telephony over the same physical pair as do ADSL and RADSL.

Seasonal Energy-efficiency Ratio (SEER). The ratio of the total seasonal cooling requirement (measured in Btu) to the total seasonal Wh of energy used, expressed in terms of Btu/Wh. (The SEER rating equals 3.413 times the seasonal COP.)

SECAM. Systeme Electronique Couleur Avec Memoire, the 625-line color television standard used in France and the former Soviet Union.

Secondary. Output winding of a transformer. Winding that is connected to a load.

Secondary Cell. Electrolytic cell used to store electricity. Once discharged may be restored by recharging by putting current through the cell in the direction opposite to that of discharge current.

Selectivity. Characteristic of a circuit to discriminate between wanted and unwanted signals.

Self Biasing. Gate bias for a field effect transistor in which source current through a resistor produces the voltage for gate to source bias.

Self Discharge. The rate at which a battery, without a load, will lose its charge.

Self Inductance. Property that causes a counterelectromotive force to be produced in a conductor when the magnetic field expands or collapses with a change of current.

Semiconductor. 1. An element which is neither a good conductor or a good insulator, but rather lies somewhere between the two. Characterized by a valence shell containing four electrons. Silicon, germanium and carbon are the semiconductors most frequently used in electronics.

2. Any material that has a limited capacity for conducting an electric current. Certain semiconductors, including silicon, gallium arsenide, copper indium dislenide, and cadmium telluride, are uniquely suited to the photovoltaic conversion process.

Semicrystalline. See 'Multicrystalline'.

Send-Only Codec. A video codec only able to originate (transmit) communications signals, for use at the sending location in point-to-multipoint or broadcast applications, and for security and military applications when two-way codec communication between locations is not possible or required.

Series Circuit. Circuit in which the components are connected end to end so that current has only one path to follow through the circuit.

Series Connection. A way of joining photovoltaic cells by connecting positive leads to negative leads; such a configuration increases the voltage.

Series Parallel Network. Network that contains components connected in both series and parallel.

Series Regulator. Type of battery charge regulator where the charging current is controlled by a switch connected in series with the PV module or array.

Series Resistance. Parasitic resistance to current flow in a cell due to mechanisms such as resistance from the bulk of the semiconductor material, metallic contacts, and interconnections.

Series Resonance. Condition that occurs in a series LC circuit at the frequency where inductive reactance equals capacitive reactance. Impedance is minimum, current is maximum limited only by resistance in the circuit.

Seven Segment Display. Device made of several light emitting diodes arranged in a numeric or alphanumeric pattern. By lighting selected segments numeric or alphabet characters can be displayed.

Shelf Life of Batteries. The length of time, under specified conditions, that a battery can be stored so that it keeps its guaranteed capacity.

Shells or Bands. Orbital path containing a group of electrons having a common energy level.

Shield. Metal grounded cover used to protect a wire, component or piece of equipment from stray magnetic and/or electric fields.

Short Circuit. Also called a short. Low resistance connection between two points in a circuit typically causing excessive current.

Short-circuit Current (Isc). The current flowing freely from a photovoltaic cell through an external circuit that has no load or resistance; the maximum current possible.

Shunt Resistor. Resistor connected in parallel or in shunt with another component or circuit.

Shunt Regulator. Type of a battery charge regulator where the charging current is controlled by a switch connected in parallel with the PV generator. Overcharging of the battery is prevented by shorting the PV generator.

Siemens Process. A commercial method of making purified silicon.

Signal. Electrical quantity that conveys information.

Signal to Noise Ratio. Ratio of the magnitude of the signal to the magnitude of noise usually expressed in decibels.

Silicon (Si). 1. Non metallic element (atomic number 14) used in pure form as a semiconductor. 2. A chemical element, atomic number 14, semimetallic in nature, dark gray, an excellent semiconductor material. A common constituent of sand and quartz (as the oxide). Crystallizes in face-centered cubic lattice like a diamond. The most common semiconductor material used in making photovoltaic devices.

Silicon-controlled Rectifier (SCR). Three terminal active device that acts as a gated diode. The gate terminal is used to turn the device on allowing current to pass from cathode to anode.

Silicon Controlled Switch. An SCR with an added terminal called an anode gate. A positive pulse either at the anode gate or the cathode gate will turn the device on.

Silicon Dioxide. Glass like material used as the gate insulating material in a MOSFET.

Silicon Transistor. A bipolar junction transistor using silicon as the semiconducting material.

Silver (Ag). Precious metal that does not easily corrode and is more conductive than copper.

Silver Mica Capacitor. Mica capacitor with silver deposited directly onto the mica sheets instead of using conductive metal foil.

Silver Solder. Solder composed of silver, copper and zinc. Has a melting point lower than pure silver, but higher than lead-tin solder.

SIM (Subscriber Identity Module). Synonymous with smart card.

Simplex. 1. Communication in only one direction at a time. Example: FAX. 2. A radio technology that allows only one-way communication. The FM radio in your car, or your TV set, could be viewed as simplex devices.

Simulcast. 1. Broadcasting a program simultaneously in two different forms, for example a program broadcast in both AM and FM. 2. A signaling technique that broadcasts the same signal over each site in a network.

Sine. Sine of an angle of a right angle triangle is equal to the opposite side divided by the hypotenuse.

Sine Wave. Wave whose amplitude is the sine of a linear function of time. It is plotted on a graph that plots amplitude against time or radial degrees relative to the angular rotation of an alternator.

Sine Wave Inverter. An inverter that produces utility-quality, sine wave power forms.

Single-crystal Material. A material that is composed of a single crystal or a few large crystals.

Single In-line Package. Package containing several electronic components (generally resistors) with a single row of connecting pins.

Single Pole Double Throw (SPDT). Three terminal switch in which one terminal can be connected to either one of the other terminals.

Single Pole Single Throw (SPST). Two terminal switch or relay that can open or close one circuit.

Single Sideband (SSB). AM radio communication technique in which the transmitter suppresses one sideband and therefore transmits only a single sideband.

Single Throw Switch. Switch containing only one set of contacts which can be either opened or closed.

Sink. Device such as a load that consumes power or conducts away heat.

Sintering. Process of bonding either a metal or powder by cold pressing it into a desired shape and then heating to form a strong cohesive body.

Sinusoidal. Varying in proportion to the sine of an angle or time function. AC voltage in which the instantaneous value is equal to the sine of the phase angle times the peak value.

SIP. 1. Abbreviation for "single in-line package." 2. *Session Initiation Protocol.* An emerging Internet standard to facilitate mixed-media sessions over IP. Sessions may span the range of services from voice-only to full multimedia conferencing; multipoint is also included.

SGML. Standard Generalized Markup Language, the superset from which HTML was derived.

Skin Effect. Tendency of high-frequency (rf) currents to flow near the surface layer of a conductor.

Slamming. The unauthorized switching of a customer's phone service to another carrier.

Sleep Mode. Designed to conserve battery life, this mode automatically turns off a terminal after it has been unused for a specified period of time. The unit is reactivated when the keypad is touched.

Slew Rate. The maximum rate at which the output voltage of an op-amp can change.

Slide Switch. Switch having a sliding button, bar or knob.

SLIP. Serial Line Interface Protocol, an Internet access method. Compare with PPP.

Slow Acting Relay. Slow operating relay that when energized may not pull up the armature for several seconds.

Slow-blow Fuse. Fuse that can withstand a heavy current (up to ten times its rated value) for a small period of time before it opens.

Smart Antenna. An antenna system whose technology enables it to focus its beam on a desired signal to reduce interference. A wireless network would employ smart antennas at its base stations in an effort to reduce the number of dropped calls, improve call quality and improve channel capacity.

Smart Card. A plastic card containing important data about a person's identity to allow access to a network or premises. Also, a card containing subscriber information, often inserted into GSM phones for roaming to different countries.

Smart Phone. A class of wireless phones typically used to describe handsets with many features and often a keyboard. What makes the phone 'smart' is its ability to handle data, not only voice calls.

SMDS. *Switched Multi-megabit Data Service,* a cell-relay wireless technology for providing data between 1.5-Mbps and 45-Mbps.

SMPTE. Society of Motion Picture and Television Engineers.

SMR (Specialized Mobile Radio). A dispatch radio and interconnect service for businesses. Covers frequencies in the 220 MHz, 800 MHz and 900 MHz bands.

SMS (Short Message Service). Electronic messages on a wireless network.

SMT (Surface Mount Technology). A surface mount device is a component, either active or passive, having no separate leads but which is part of the component body to permit direct mounting on a printed circuit board.

SNA. System Network Architecture (IBM usage).

Snap Switch. Switch containing a spring under tension or compression that causes the contacts to come together suddenly when activated.

SNMP. Simple Network Management Protocol (TCP/IP).

SNR. Abbreviation for "signal to noise ratio."

Soft Handoff. Procedure in which two base stations-one in the cell site where the phone is located and the other in the cell site to which the conversation is being passed- both hold onto the call until the handoff is completed. The first cell site does not cut off the conversation until it receives information that the second is maintaining the call.

Soft Key. A key below the phone's display that performs whatever function is listed on the display.

Soft Magnetic Material. Ferromagnetic material that is easily demagnetized.

Softswitch. A concept that was perhaps first formalized by the ITU-T Working Party developing the nascent T.130 series recommendations in the mid-1990s. Several key concepts included distributed network intelligence, distributed switching (time and space), and the separation of the multimedia content and the switching logic layers.

Now, there is an industry consortium promulgating standards-based aspects including:

— H.323 videoconferencing and multimedia

— SIP Session Initiation Protocol

— RTP Real-time Transport Protocol

— RTSP Real-Time Streaming Protocol

— MGCP Media Gateway Control Protocol

Depending upon the point of view of the particular industry sector, Softswitch technology may result in the replacement of the end-office switch for the CLEC for the provision of telephony and facsimile. At the other extreme, Softswitch can be a flexible, intelligent packet switching arrangement in a mixed multimedia environment. Internet services like VoIP will also likely be part of this technology mix, once the quality of service issues are fully resolved.

Software. Program of instructions that directs the operation of a computer.

Solar Cell. Photovoltaic cell that converts light into electric energy. Especially useful as a power source for space vehicles.

Solar Constant. The strength of sunlight; 1353 watts per square meter in space and about 1000 watts per square meter at sea level at the equator at solar noon.

Solar Energy. Energy from the sun. The heat that builds up in your car when it is parked in the sun is an example of solar energy.

Solar-grade Silicon. Intermediate-grade silicon used in the manufacture of solar cells. Less expensive than electronic-grade silicon.

Solar Noon. That moment of the day that divides the daylight hours for that day exactly in half. To determine solar noon, calculate the length of the day from the time of sunset and sunrise and divide by two. Solar noon may be quite a bit different from 'clock' noon.

Solar Spectrum. The total distribution of electromagnetic radiation emanating from the sun.

Solar Thermal Electric. Method of producing electricity from solar energy by using focused sunlight to heat a working fluid, which in turn drives a turbogenerator.

Solder. Metallic alloy used to join two metal surfaces.

Soldering. Process of joining two metallic surfaces to make an electrical contact by melting solder (usually tin and lead) across them.

Soldering Iron. Tool with an internal heating element used to heat surfaces being soldered to the point where the solder becomes molten.

Solenoid. An air core coil. Equipped with a movable iron core the solenoid will produce motion. As a result of current through the coil the iron core is pulled into the center of the winding. When the coil is deenergized, a spring pulls the movable core away from the center of the winding. Mechanical devices connected to the movable core are made to move as a result of current through the coil. Example: Electric door locks on some automobiles.

Solid Conductor. Conductor having a single solid wire instead of strands of fine wire twisted together.

Solid State. Pertaining to circuits where signals pass through solid semiconductor material such as transistors and diodes as opposed to vacuum tubes where signals pass through a vacuum.

Sonar. Acronym for "sound navigation and ranging." A system using reflected sound waves to determine the position of some target.

SONET. Synchronous Optical NETwork.

Sonic. Pertaining to sound.

Sound Wave. Pressure waves propagated through air or other plastic media. Sound waves are generally audible to the human ear if the frequency is between approximately 20 and 20,000 vibrations per second (hertz).

Source. Device that provides signal power or energy to a load.

Source Follower. FET amplifier in which signal is applied between gate and drain with output taken between source and drain. Also called "common drain."

Source Impedance. Impedance through which output current is taken from a source.

South Pole. Pole of a magnet into which magnetic lines of force are assumed to enter.

Space Charge. See 'cell barrier'; 'depletion zone.

Spark. Momentary discharge of electrical energy due to ionization of air or other dialectric material separating two charges.

Spatial Filtering. In video compression, removal of redundant information on a line-by-line basis by discarding diagonally adjacent pixels. Also known as *field subsampling, loop filtering,* H.261, H.263 usage.

SPDT. Single pole double throw.

Speaker. Also called "loudspeaker." Transducer that converts electrical energy into mechanical energy at audio frequencies.

Spectrum. Arrangement or display of light or other forms of electromagnetic radiation separated according to wavelength, energy or some other property.

Spectrum Allocation. Federal government designation of a range of frequencies for a category of use or uses. For example, the FCC allocated the 1900 MHz band for personal communications services. Allocation, typically accomplished in years-long FCC proceedings, tracks new technology development. However, the FCC can shift existing allocations to accommodate changes in spectrum demand. As an example, some UHF television channels were recently reallocated to public safety.

Spectrum Analyzer. Instrument used to display the frequency domain of a waveform plotting amplitude against frequency.

Spectrum Assignment. Federal government authorization for use of specific frequencies or frequency pairs within a given allocation, usually at stated a geographic location(s). Mobile communications authorizations are typically granted to private users, such as oil companies, or to common carriers, such as cellular and paging operators. Spectrum auctions and/or frequency coordination processes, which consider potential interference to existing users, may apply.

Spectrum Cap. A limit to the allocated spectrum designated for a specific service.

Spectrum Etiquette. Scheme under which various brands of equipment for unlicensed-band communications can share the same frequencies. For example, a 'listen-before-talk' etiquette would have all devices first sense if a channel is clear.

Speed-up Capacitor. Capacitor added to the base circuit of a BJT switching circuit to improve the switching time of the device.

Spinning Reserve. Utility generating capacity on-line and running at low power in excess of actual load.

SPL. Sound Pressure Level, in acoustic measurement.

Split-spectrum Cell. A compound photovoltaic device in which sunlight is first divided into spectral regions by optical means. Each region is then directed to a different photovoltaic cell optimized for converting that portion of the spectrum into electricity. Such a device achieves significantly greater overall conversion of incident sunlight into electricity. See 'mulitjunction device'.

Spread Spectrum. Jamming-resistant and initially devised for military use, this radio transmission technology 'spreads' information over greater bandwidth than necessary for interference tolerance and is now a commercial technology.

SPST. Abbreviation for "single pole single throw."

Sputtering. A physical vapor deposition process where high-energy ions are used to bombard elemental sources of semiconductor material, which eject vapors of atoms that are then deposited in thin layers on a substrate.

SQCIF. Sub-Quarter Common Intermediate Format. Describes a subset of the type of coded video signals transmitted when using ITU-T Rec. H.261 and H.263 coding methods. See *CIF.*

Square Wave. Wave that alternates between two fixed values for an equal amount of time.

Square Wave Inverter. The inverter consists of a dc source, four switches, and the load. The switches are power semiconductors that can carry a large current and withstand a high voltage rating. The switches are turned on and off at a correct sequence, at a certain frequency. The square wave inverter is the simplest and the least expensive to purchase, but it produces the lowest quality of power.

SRAM (Static Random Access Memory). A memory technology used in pagers and handsets. So named because it requires no refresh cycle, as required by dynamic RAM (DRAM) and therefore consumes less power. SRAM maintains data only while power is applied.

SS7 (Signaling System 7). An international high speed signaling backbone for the public switched telephone network.

SSFH. Spread Spectrum Frequency Hopping, used for data radio transmission from the low 10's of MHz to the multi-GHz bands, especially under high interference conditions. Popular for wireless LANs, for example

Staebler-Wronski Effect. The tendency of amorphous silicon photovoltaic devices to lose efficiency upon initial exposure to light; named for Dr. David Staebler and Dr. Christopher Wronski; work performed at RCA.

Stand-alone (PV system). An autonomous or hybrid photovoltaic system not connected to a grid. May or may not have storage, but most stand-alone systems require batteries or some other form of storage.

Standby Time. The amount of time a subscriber can leave a fully charged handset turned on to receive incoming calls before the phone will discharge the batteries.

Stand-off Mounting. Technique for mounting a PV array on a sloped roof, which involves mounting the modules a short distance above the pitched roof and tilting them to the optimum angle.

Standard Reporting Conditions (SRC). A fixed set of conditions (including meteorological) to which the electrical performance data of a photovoltaic module are translated from the set of actual test conditions. [ASTM E 1036]

Standard Test Conditions (STC). Conditions under which a module is typically tested in a laboratory: (1) Irradiance intensity of 1000 W/square meter (0.645 watts per square inch), AM1.5 solar reference spectrum, and (3) a cell (module) temperature of 25 degrees C, plus or minus 2 degrees C (77 degrees F, plus or minus 3.6 degrees F). [IEC 1215]

State of Charge (SOC). The available capacity remaining in the battery, expressed as a percentage of the rated capacity.

Static. Crackling noise heard on AM radio receivers. Caused by electric storms or electric devices.

Static Electricity. Stationary electric charges.

Static Reverse Current. Reverse current through a zener diode when the reverse voltage across the diode is less than the zener voltage rating of the device.

Stator. Stationary part of some rotary device such as a variable capacitor.

STDM. Statistical Time-Division Multiplexer, (statmux or statistical multiplexer), a device which combines a number of time-varying bit streams into a single bit stream for transmission. In DBS, may dynamically allocate bandwidth through control signals to MPEG2 video codecs.

Step-down Transformer. Transformer in which the output AC voltage is less than the input AC voltage.

Step-up Transformer. Transformer in which the output AC voltage is greater than the input AC voltage.

Stereo Sound. System in which reproduced sound is delivered through two or more channels to give a sense of direction to the source.

Stop Band. Range of frequencies outside the pass band of a tuned amplifier.

Storage Time. In a BJT switching circuit, it is the time required for collector current to drop from 100% to 90% of its maximum value.

Stranded Conductor. Conductor composed of a group of strands of wire twisted together.

Stratospheric Platform. Blimp-like platform for wireless telephone service in urban areas.

Stray Capacitance. Undesirable capacitance that exists between two conductors such as two leads or one lead and a metal chassis.

Strongest Signal. The concept that a wireless 911 call should be routed to the cell site with the strongest link to the phone, regardless of which carrier holds the caller as a customer. The strength of the call's setup link isn't always equal to that of the link the cell assigns for voice traffic; the latter can be weaker.

Subassembly. Components contained in a unit for convenience in assembling or servicing equipment.

Subatomic. Particles such as electrons, protons and neutrons that are smaller than atoms.

Subscriber Profiling. The process of compiling subscriber usage information (such as frequency of calls, locations called to or from and monthly airtime usage), typically to identify potentially fraudulent use or to identify customers likely to terminate service.

Substrate. 1. Mechanical insulating support upon which a device is fabricated. 2. The physical material upon which a photovoltaic cell is made.

Sulfation. A condition that afflicts unused and discharged batteries; large crystals of lead sulfate grow on the plate, instead of the usual tiny crystals, making the battery extremely difficult to recharge.

Summing Amplifier. An op-amp circuit whose output is proportional to the sum of its instantaneous voltages.

Sunspace. A room that faces south, or a small structure attached to the south side of a house.

Sun Tempering. A sun-tempered building is elongated in the east-west direction, with the majority of the windows on the south side. The area of the windows is generally limited to about 7% of the total floor area. A sun-tempered design has no added thermal mass beyond what is already in the framing, wall board, and so on. Insulation levels are generally high.

Superconducting Magnetic Energy Storage (SMES). SMES technology uses the superconducting characteristics of low-temperature materials to produce intense magnetic fields to store energy. SMES has been proposed as a storage option to support large-scale use of photovoltaics and wind as a means to smooth out fluctuations in power generation.

Superconductivity. The pairing of electrons in certain materials when cooled below a critical temperature, causing the material to lose all resistance to electricity flow. Superconductors can carry electric current without any energy losses.

Superconductor. Metal such as lead or niobium that, when cooled to within a few degrees of absolute zero, can conduct current with no resistance.

Superheterodyne Receiver. Radio receiver that converts all radio frequencies to a fixed intermediate frequency to maximize gain and bandwidth before demodulation.

Super High Frequency (SHF). Frequency band between 3 GHz and 30 GHz. So desiganted by Federal Communications Commission (FCC).

Superposition Theorem. Theorem designed to simplify networks containing two or more sources. It states that in a network containing more than one source, the current at any one point is equal to the algebraic sum of the currents produced by each source acting separately.

Superstrate. The covering on the sun side of a PV module, providing protection for the PV materials from impact and environmental degradation while allowing maximum transmission of the appropriate wavelengths of the solar spectrum.

Supply Voltage. Voltage provided by a power source.

Surface-barrier Diode. (Schottky diode). High speed diode that has very little junction capacitance. Also known as a "hot-carrier diode."

Surface Leakage Current. Diode reverse current that passes along the surface of the semiconductor materials.

Surge Current. High charging current that flows into a power supply filter capacitor as the power is first turned on.

SVC. Switched Virtual Circuit (frame relay use).

Sweep Generator. Test instrument designed to produce a voltage that continuously varies in frequency over a band of frequencies. Used as a source to display frequency response of a circuit on an oscilloscope.

Switch. Electrical device having two states, on (closed) or off (open). Ideally having zero impedance when closed and infinite impedance when open.

Switching Transistor. transistor designed to change rapidly between saturation and cut-off.

Symbian. The joint venture between Ericsson Inc., Motorola Inc., Nokia Corp. and Psion to develop new operating systems based on Psion's EPOC32 platform for small mobile devices including wireless phones or handheld personal computers.

Synchronization. Also called sync. Precise matching of two waves or functions.

Synchronous. With reference to video and data signals and devices: being precisely in step, or happening together in time.

Sync Pulse. Pulse used as a reference for synchronization.

Syndrome. A code transmitted along with data to enable correction of transmission-induced errors in the received data. See *Forward Error Correction*. Compare with *Checksum*.

System. Combination of several pieces of equipment to perform in a particular manner.

System Identification Number (SID). A unique number assigned to every wireless operator in the United States that is then programmed into the phones that subscriber's to that service purchase.

T

T1. The AT&T Bell System level 1 digital transmission system operating at 1.544 Mbps (1.536 Mbps excluding framing). T1 is commonly used to refer to *DS1*.

T3. The AT&T Bell System level 3 digital transmission system operating at 44.736 Mbps (commonly referred to as 45 Mbps). T3 is commonly used to refer to *DS3*.

T.120 (suite specification). An ITU-T overview recommendation for multimedia conferencing describing Transmission Protocols for Multimedia Data; including multilayer protocols to considerably enhance multimedia, MCU and codec control capabilities, permitting greater MCU operational sophistication beyond that described in H.231 and H.243.

T.121. ITU-T recommendation describing a generic application template.

T.122. ITU-T recommendation describing a multipoint communication service (MCS) for audiographic and audiovisual conferencing — service definition document.

T.123. ITU-T recommendation describing network-specific data protocol stacks for multimedia teleconference applications.

T.124. ITU-T recommendation describing Generic Conference Control for audio-visual and audiographic terminals.

T.125. ITU-T recommendation describing a multipoint communication service protocol specification.

T.126. ITU-T recommendation describing multipoint still image [Conferencing] and annotation protocol specification.

T.127. ITU-T recommendation describing multipoint binary file transfer protocol specification.

T.128. ITU-T recommendation describing multipoint application sharing specifications.

T.130 (Suite description). ITU-T T.13x recommendations describe real-time architecture for multimedia conferencing. This suite of standards introduces such concepts as virtual networks, virtual and distributed MCUs and many more innovative new network architecture approaches for voice, video, graphics, data etc. Fully supports T.120 suite and a variety of coding and transmission methods.

T.131 ?. ITU-T recommendation describing Real-Time Stream Multipoint Communications Service.

T.132 ?. ITU-T recommendation describing Real-time Link Management.

T.133 ?. ITU-T recommendation describing an Audio-Video Control Applications Protocol.

T.134. ITU-T recommendation describing a text chat application entity.

T.135. ITU-T recommendation describing user to reservation system transactions within T.120 conferences.

T.136. ITU-T recommendation describing a remote device control application protocol.

T.137. ITU-T recommendation describing virtual meeting room management – services and protocol.

T.140. ITU-T recommendation describing a protocol for multimedia application text conversation.

Tail Time. In acoustic echo cancellers, telephony echo cancellers and related devices, the sampling window size (in milliseconds) used for the electroacoustic model of all the echo sources. Related to the more complex AFIRF "taps" parameter.

Tank Circuit. Parallel resonant circuit containing only a coil and a capacitor. Both the coil and capacitor store electrical energy for part of each cycle.

Tantalum Capacitor. Electrolytic capacitor having a tantalum foil anode. Able to have a large capacity in a small package.

Tap. Electrical connection to some point other than at the ends of a resistor or inductor.

Tapered. Nonunifrom distribution of resistance per unit length throughout the element of a potentiometer.

TCP/IP. *Transmission Control Protocol/Internet Protocol.* Networking protocols providing communications across interconnected networks (including Internet and Intranets). The networks may incorporate diverse hardware architecture computers and various operating systems. TCP/IP includes communication, connection and routing standard protocols.

TDD. *Time Division Duplex*, part of DECT modulation scheme.

TDM. *Time-Division Multiplex*, the sharing of a transmission network by a number of users versus time.

TDMA (Time Division Multiple Access). 1. A digital air interface technology used in cellular, PCS and ESMR networks. 2. *Time-Division Multiple Access*, the sharing of a satellite transmission network by a number of locations versus time.

TEC. Telephony Echo Canceller, an echo canceller used to eliminate echoes on the line side of AECs and other telecommunications voice equipment.

Technician. Expert in troubleshooting circuit and system malfunctions. Along with a thorough knowledge of test equipment and how to use it to diagnose problems, the technician is also familiar with how to repair or replace faulty components. Technicians basically translate theory into action.

Telco. Jargon for TELephone COmpany.

Telegraphy. Communication between two points by sending and receiving a series of current pulses either through wire or by radio.

Telematics. The integration of wireless communications, vehicle monitoring systems and location devices.

Telemetry. Transmission of instrument readings to a remote location either by wire or by radio.

Telephone. Apparatus designed to convert sound waves into electrical waves which are sent to and reproduced at a distant point.

Telephone Line. Wires existing between subscribers and central stations in a telephone system.

Telephony. Telecommunications system involving the transmission of speech information, allowing two or more persons to communicate verbally.

Teletypewriter. Electric typewriter that like a teleprinter can produce coded signals corresponding to the keys pressed or print characters corresponding to the coded signals received.

Television. System that converts both audio and visual information into corresponding electrical signals which are then transmitted through wires or by radio waves to a receiver which reproduces the original information.

Telex. Teletypewriter exchange service.

Temperature Coefficient of Frequency. Rate at which frequency changes with temperature.

Temporal Filtering. In video compression, the recursive removal of redundant pixels on a frame-to-frame basis, a major factor controlling final image quality.

Tera (T). Metric prefix that represents 10^{12}.

Terminal. Point at which electrical connections are made.

Termination Charges. Fees that wireless telephone companies pay to complete calls on wireline phone networks or vice versa.

Tesla (T). Unit of magnetic flux density. (1 tesla = 1 Wb/m^2).

Test. Sequence of operations intended to verify the correct operation or malfunctioning of a piece of equipment or system.

TETRA (Terrestrial Trunked Radio). An open digital trunked radio standard defined by the European Telecommunications Standardization Institute.

Thermal Electric. Electric energy derived from heat energy, usually by heating a working fluid, which drives a turbogenerator. See 'solar thermal electric'.

Thermal Mass. Materials that store heat within a sunspace or solar collector.

Thermal Relay. Relay activated by a heating element.

Thermal Runaway. Problem that can develop in an amplifier when an increase in temperature causes an increase in collector current. The increase in collector current causes a further increase in temperature and so on. Unless the circuit is designed to prevent this condition, the device can be driven into saturation.

Thermal Stability. The ability of a circuit to maintain stable characteristics in spite of increased temperature.

Thermal Storage Walls (masonry or water). A thermal storage wall is a south-facing wall that is glazed on the outside. Solar heat strikes the glazing and is absorbed into the wall, which conducts the heat into the room over time. The walls are at least 8 in thick. Generally, the thicker the wall, the less the indoor temperature fluctuates.

Thermistor. Temperature sensitive semiconductor that has a negative temperature coefficient of resistance. As temperature increases, resistance decreases.

Thermocouple. Temperature transducer consisting of two dissimilar metals welded together at one end to form a junction that when heated will generate a voltage.

Thermometry. Relating to the measuring of temperature.

Thermophotovoltaic (TPV) Device. A device that converts secondary thermal radiation, re-emitted by an absorber or heat source, into electricity; The device is designed for maximum efficiency at the wavelength of the secondary radiation.

Thermostat. Device that opens or closes a circuit in response to changes in temperature.

Thevenin's Theorem. Theorem that replaces any complex network with a single voltage source in series with a single resistance.

Thick-crystalline Materials. Semiconductor material, typically measuring from 200-400 microns thick, that is cut from ingots or ribbons.

Thick-film Capacitor. Capacitor consisting of two thick-film layers of conductive film separated by a deposited thick-layer dialectric film.

Thick Film Resistor. Fixed value resistor consisting of thick-film resistive element made from metal particles and glass powder.

Thin Client. A pen-based tablet computer used on a wireless local area network.

Thin Film. A layer of semiconductor material, such as copper indium diselenide, cadmium telluride, gallium arsenide, or amorphous silicon, a few microns or less in thickness, used to make photovoltaic cells.

Thin Film Capacitor. Capacitor in which both the electrodes and the dialectric are deposited in layers on a substrate.

Thin Film Detector (TFD). A temperature detector containing a thin layer of platinum and used for precise temperature readings.

Third-generation. A new standard that promises to offer increased capacity and high-speed data applications up to 2 megabits. It also will integrate pico-, micro- and macrocellular technology and allow global roaming. (Also see '3G.')

III-V (three-five) Materials. Elemental materials that occupy groups III and V of the Periodic Table of the Elements.

3G (Third Generation). The next generation of wireless technology beyond personal communications services. The World Administrative Radio Conference assigned 230 megahertz of spectrum at 2 GHz for multimedia 3G networks. These networks must be able to transmit wireless data at 144 kilobits per second at mobile user speeds, 384 kbps at pedestrian user speeds and 2 megabits per second in fixed locations. The International Telecommunication Union seeks to coordinate 3G standards through its International Mobile Telecommunications-2000 project. In early July, the ITU received 10 proposals for 3G systems and is currently holding a series of meetings to evaluate the specifications.

Thread-of-Life Communications. Emergency telecommunications facilities, generally of a substandard nature and where only minimal communications facilities or bandwidth are available. For example, videoconferencing at 19.2 Kbps will provide visual identification of a talker under wartime conditions in the bandwidth equivalent to one-third of a regular G.711 telephone voice circuit.

Three Phase Supply. AC supply that consists of three AC voltages 120□ out of phase with each other.

Threshold. Minimum point at which an effect is produced or detected.

Threshold Voltage. For an enhancement MOSFET, the minimum gate source voltage required for conduction of source drain current.

Thyristor. A term used to classify all four layer semiconductor devices. SCRs and triacs are examples of thyristors.

TIA (Telecommunications Industry Association). A trade group representing manufacturers and suppliers of communications and information technology products. TIA is a standards-developing organization accredited by the American National Standards Institute.

Timebase Corrector (TBC). A time domain jitter removal device used in conjunction with a playback from a VCR. It provides sufficient timing stability to permit successful encoding of the video signal by a video codec. Some codecs include a built-in TBC.

Time Constant (t). Time required for a capacitor in an RC circuit to charge to 63% of the remaining potential across the circuit. Also time required for current to reach 63% of maximum value in an RL circuit. Time constant of an RC circuit is the product of R and C. Time constant of an RL circuit is equal to inductance divided by resistance.

Time Division Multiplexing (TDM). Transmission of two or more signals on the same path, but at different times.

Time-domain Analysis. A method of representing a waveform by plotting amplitude over time.

Tin Oxide. A wide band-gap semiconductor similar to indium oxide; used in heterojunction solar cells or to make a transparent conductive film, called NESA glass when deposited on glass.

Toggle Switch. Spring loaded switch that is put in one of two positions either on or off.

Tolerance. Permissable deviation from a specified value normally expressed as a percentage.

TO Package. Cylindrical, metal can type of package of some semiconductor components.

Toroidal Coil. Coil wound on a doughnut shaped core.

Total Energy System. See 'photovoltaic-thermal system'.

Total Harmonic Distortion (thd). The measure of closeness in shape between a waveform and its fundamental component.

Total Internal Reflection. The trapping of light by refraction and reflection at critical angles inside a semiconductor device so that it cannot escape the device and must eventually be absorbed by the semiconductor.

TQM, Total Quality Management. A corporate philosophy or mindset focused on quality, a precursor of ISO 9000.

Tracking PV Array. PV array that follows the path of the sun to maximize the solar radiation incident on the PV surface. The two most common orientations are (1) one axis where the array tracks the sun east to west and (2) two-axis tracking where the array points directly at the sun at all times. Tracking arrays use both the direct and diffuse sunlight. Two-axis tracking arrays capture the maximum possible daily energy.

Transcoder. Hardware or software that converts between algorithms, protocols or standards.

Transconductance. Also called mutual conductance. Ratio of a change in output current to the change in input voltage that caused it.

Transducer. Device that converts energy from one form to another.

Transformer. 1. Inductor with two or more windings. Through mutual inductance, current in one winding called a primary will induce current into the other windings called secondaries.

2. Converts the generator's low-voltage electricity to higher voltage levels for transmission to the load center, such as a city or factory.

Transformer Coupling. Also called inductive coupling. Coupling of two circuits by means of mutual inductance provided by a transformer.

Transistor. Term derived from "transfer resistor." Semiconductor device that can be used as an amplifier or as an electronic switch.

Transmission. Sending of information.

Transmission Line. Conducting line used to transmit signal energy between two points.

Transmission Lines. Transmit high-voltage electricity from the transformer to the electric distribution system.

Transmitter. Equipment used to achieve transmission.

Transponder (satellite usage). Part of a communications satellite that receives a signal from the earth, amplifies it, converts it to a lower frequency and re-transmits it back to the earth. Each transponder is essentially a broad-band amplifier with a bandwidth of about 36 MHz and a transmit power ranging from a few watts to hundreds of watts depending upon the application. Each transponder can carry thousands of telephone conversations or one to many television channels. Satellites include a number of transponders with 16 to 48 being typical of different applications.

Triac. Bidirectional gate controlled thyristor similar to an SCR, but capable of conducting in both directions. Provides full wave control of AC power.

Triangular Wave. A repeating wave that has equal positive going and negative going ramps. The ramps have linear rates of change with time.

Triangulation. The lengthy process of pinning down a caller's location using radio receivers, a compass and a map.

Trickle Charge. A charge at a low rate, balancing through self-discharge losses, to maintain a cell or battery in a fully charged condition.

Trigger. Pulse used to initiate a circuit action.

Triggering. Initiation of an action in a circuit which then functions for a predetermined time. Example: The duration of one sweep in a cathode ray tube.

Trimmer. Small value variable capacitor, resistor or inductor used to fine tune a larger value.

Tri-Mode. The term applied to a phone that will work on 800 MHz analog, 800 MHz digital and 1900 MHz (also known as 1.9 GHz) frequencies.

Tri-mode Handset. Phones that work on three frequencies, typically using 1900 MHz, 800 MHz digital or reverting to 800 MHz analog cellular when digital is not available.

Triple Band. A network infrastructure or wireless phone designed to operate in three frequency bands.

Trivalent Element. One having three valence electrons. Used as an impurity in semiconductor material to produce p-type material. Most commonly used trivalent elements are: Aluminum, Gallium and Boron.

Troubleshooting. Systematic approach to locating the cause of a fault in an electronic circuit or system.

Trunking. Spectrum-efficient technology that establishes a queue to handle demand for voice or data channels.

TTL. Transistor-Transistor Logic. A type of integrated circuit with nominal logic levels of +5 and 0 volts. These logic levels are often specified for digital data and control signals that connect codecs and other communications equipment.

Tumbling. The process of changing the Electronic Serial Number (ESN) in a single phone each time that a phone call is made, thus allowing the user to make calls and have then illegally charged to someone else's number.

Tuned Circuit. Circuit that can have its component values adjusted so that it responds to one selected frequency and rejects all others.

Tunnel Diode. Heavily doped junction diode that has negative resistance in the forward direction of its operating range.

Tunneling. Quantum mechanical concept whereby an electron is found on the opposite side of an insulating barrier without having passed through or around the barrier.

Turnkey. an entire system with hardware and software assembled and installed by a vendor and sold as a total package.

Turn-off Time. Sum of storage time and fall time.

Turn-on Time. Sum of delay time and rise time.

Turns Ratio. Ratio of the number of turns in the secondary winding of a transformer to the number of turns in the primary winding.

TVRO. *TeleVision Receive-Only* satellite earth station.

Two-axis Tracking. A system capable of rotating independently about two axes (e.g., vertical and horizontal).

Two Phase. Two repeating waveforms having a phase difference of 90.

U

UHF. 1. Abbreviation for "ultra high frequency. 2. Referring to radio channels in the 300 MHz to 3 GHz band.

ULS (Universal Licensing System). The new Wireless Telecommunications Bureau program under which electronic filing of license applications and reports of changes to licenses creates a database that can be accessed remotely for searches. Using ULS, for example, the user can learn all the specialized mobile radio licenses in a given region.

Ultrasonic. Signals that are just above the frequency range of human hearing of approximately 20 kHz.

Ultraviolet. Electromagnetic radiation in the wavelength range of 4 to 400 nanometers. vacuum evaporation-The deposition of thin films of semiconductor material by the evaporation of elemental sources in a vacuum.

UMTS (Universal Mobile Telecommunications System). Europe's approach to standardization for third-generation cellular systems.

Uncharged. Material having atoms with the same number of electrons in orbit as the number of protons in the nucleus. Having no electrical charge.

Unified Messaging. Software technology that allows carriers and Internet service providers to manage customer e-mail, voice and fax messages from any phone, PC or information device.

Unijunction Transistor. Three terminal device that acts as a diode with its own internal voltage divider biasing circuit.

Unity Gain Frequency. Frequency of operation for a device where the gain of the component drops to unity.

Universal Service. The government's aim, starting in the 1930s, of providing phone service to all, regardless of distance from the switch or ability to pay. Today, universal service encompasses those aims, plus a subsidy to public schools, libraries and rural health care facilities for telecom services.

Unvalidated Phone. A wireless phone lacking a service contract. Also called an unregistered phone.

Uplink. The portion of a telecommunications path from the ground to the satellite. Also referred to as the forward link.

UPS. *Uninterruptible Power Supply*. A secondary power supply that provides service when the main AC power degrades or fails. A UPS may include batteries or a motor generator (MG) set with flywheel energy storage (a no-break system).

USAT (Ultra Small Aperture Terminal). Satellite receive dishes for telemetry and other remote monitoring, usually smaller than VSATs.

Utility-interactive Inverter. An inverter that can function only when tied to the utility grid, and uses the prevailing line-voltage frequency on the utility line as a control parameter to ensure that the PV system's output is fully synchronized with the utility power.

UWC-136. A third-generation wireless standard proposal based on TDMA technology that was developed by the Universal Wireless Communications Consortium and is one of the 3G candidates submitted to the International Telecommunication Union by the United States.

UWCC (Universal Wireless Communications Consortium). An industry group supporting IS-136 time division multiple access and IS-41 wireless intelligent network technology.

V

V.11. ITU-T rec. describing electrical characteristics for balanced double-current interchange circuits operating at data signaling rates up to 10 Mbps.

V.24. ITU-T rec. listing definitions for interchange circuits between terminal equipment (DTE) and data circuit-terminating equipment (DCE).

V. Vertical, Volts, Velocity.

VA. Abbreviation for "volt ampere".

Vac. Volts ac.

Vacuum Evaporation. Method of depositing thin coatings of a substance by heating it in a vacuum system.

Vacuum Tube. Electron tube evacuated to such a degree that its electrical characteristics are essentially unaffected by the presence of residual gas or vapor. Have been essentially replaced by transistors for amplification and rectification. Cathode ray tubes are still used as display devices.

Vacuum Zero. The energy of an electron at rest in empty space; used as a reference level in energy band diagrams.

Valence Band. The highest energy band in a semiconductor that can be filled with electrons.

Valence Shell. The outermost electron shell for a given atom. The number of electrons in this shell determines the conductivity of the atom.

Valence State; Valence Level Energy, Bound State. Energy content of an electron in orbit about an atomic nucleus.

VAPC. Vector Adaptive Predictive Coding. A proprietary audio coding algorithm primarily intended for use at the lowest transmission aggregate bit rates.

Varactor Diode. PN junction diode with a high junction capacitance when reverse biased. Most often used as a voltage controlled capacitor. The varactor is also called: varicap, tuning diode and epicap.

Variable Capacitor. Capacitor whose capacitance can be change by varying the effective area of the plates or the distance between the plates.

Variable Resistor. Resistor whose resistance can be changed by turning a shaft. See also "potentiometer and rheostat."

VCR. Abbreviation for "video cassette recorder."

VCXO (Voltage-controlled Crystal Oscillator). A crystal oscillator is an oscillator in which the frequency is controlled by a piezoelectric crystal. Types of crystal oscillators include voltage-controlled crystal oscillators (VCXO), temperature-compensated crystal oscillators (TCXO), oven-controlled crystal oscillators (OCXO), temperature-compensated-voltage controlled crystal oscillators (TCVCXO), oven-controlled voltage-controlled crystal oscillators (OCVCXO), microcomputer-compensated crystal oscillators (MCXO), and rubidium crystal oscillators (RbXO).

Vdc. Volts dc.

VDSL. *Video (or Very-high bit rate) Digital Subscriber Line*, the fastest form of xDSL, it provides asymmetrical full duplex connectivity at 52 Mbps to the subscriber and 16 Mbps from the subscriber, over distances of up to 1.2 Km (0.75 miles) over a single pair. This type of transmission does not support telephony over the same physical pair as do ADSL and RADSL.

Vector. Quantity having both magnitude and direction. Normally represented by a line. Length of the line indicates magnitude and orientation indicates direction.

Vector Diagram. Arrangement of vectors showing phase relationships between two or more AC quantities of the same frequency.

Vertical MOS. Enhancement type MOSFET designed to handle much greater values of drain current than standard E-MOSFET.

Vertical Multijunction (VMJ) Cell. A compound cell made of different semiconductor materials in layers, one above the other. Sunlight entering the top passes through successive cell barriers, each of which converts a separate portion of the spectrum into electricity, thus achieving greater total conversion efficiency of the incident light. Also called a multiple junction cell. See 'multijunction device'; 'split-spectrum cell'.

Very High Frequency (VHF). Electromagnetic frequency band from 30 MHz to 300 MHz.

Very Low Frequency (VLF). Frequency band from 3 kHz to 30 kHz.

Video. Relating to any picture or visual information. From the Latin word meaning "I see."

Video Amplifier. Amplifier having one or mare stages designed to amplify video signals.

Video Server. An audio and video file server, often with multiple terabyte hard disk capacities, connected to LANs and WANs with high-speed facilities like ATM or FDDI. Depending on use, may employ H.320, H.323, MPEG2, MPEG4 or Motion-JPEG video compression for broadcasting, training on demand to the desktop and many other corporate purposes including marketing. Also a key technology for VoD (which see).

VIRIS. A BellCore-developed VIdeo Reference Impairment System, a subset of RISV, useful for determining acceptable impairment levels for compression schemes including MPEG1, MPEG2 and MPEG4, H.320, H.323 and others. VIRIS may be used as the preparatory phase of a P.910 subjective testing program.

Virtual Ground. Point in a circuit that is always at approximately ground potential. Often a ground for voltage, but not for current.

VLR (Visitor Location Register). A network database that holds information about roaming customers.

Vmp. Voltage at maximum power.

Voc. Open-circuit voltage.

VoD. *Video-on-Demand*, a combination of video servers, networks, infrastructure, supporting software and user interface to provide broadcast, pay-TV, training or reference data on demand. Some applications also use IRD hardware as a user terminal.

VoDSL. Voice over Digital Subscriber Loop, techniques permitting transmission of telephony over DSL.

Voice Activation. A feature that allows a subscriber to dial a phone by spoken commands.

Voice Coil. Coil attached to the diaphram of a moving coil loudspeaker. The coil is moved through an air gap between magnetic pole pieces.

Voice Recognition. The capability for cellular phones, PCs and other communications devices to be activated or controlled by voice commands.

Voice Synthesizer. Synthesizer that can simulate speech by stringing together phenomes.

VoIP. Voice over Internet Protocol, a series of techniques permitting transmission of telephony over the Internet. Often makes use of ITU-T G.7xx audio compression recommendations.

Volt. 1. Unit of potential difference or electromotive force. One volt is the potential difference needed to produce one ampere of current through a resistance of one ohm.

2. A unit of measure of the force, or 'push,' given the electrons in an electric circuit. One volt produces one ampere of current when acting a resistance of one ohm.

Voltage (V). Term used to designate electrical pressure or force that causes current to flow.

Voltage Amplifier. Amplifier designed to build up signal voltage. By design amplifiers can have a large voltage gain or a large current gain or a large power gain. Voltage amplifiers are designed to maximize voltage gain often at the expense of current gain or power gain.

Voltage at Maximum Power (Vmp). The voltage at which maximum power is available from a module. [UL 1703]

Voltage Controlled Oscillator. Oscillator whose output frequency depends on an input control voltage.

Voltage Divider. Fixed or variable series resistor network connected across a voltage to obtain a desired fraction of that voltage.

Voltage Divider Biasing. Biasing method used with amplifiers in which two series resistors connected across a source. The junction of the two biasing resistors provides correct bias voltage for the amplifier.

Voltage Drop. Voltage or difference in potential developed across a component due to current flow.

Voltage Feedback. Feedback configuration where a portion of the output voltage is fed back to the input of an amplifier.

Voltage Follower. Operational amplifier circuit characterized by a high input impedance, low output impedance and unity voltage gain. Used as a buffer between a source and a low impedance load.

Voltage Gain. Also called voltage amplification. Ratio of amplifier output voltage to input voltage usually expressed in decibel

Voltage Multiplier. Rectifier circuit using diodes and capacitors to produce a DC output voltage that is some multiple of the peak value of AC input voltage. Cost effective way of producing higher DC voltages. Voltage doublers and voltage triplers are examples.

Voltage Rating. Maximum voltage a component can withstand without breaking down.

Voltage Regulator. Device or circuit that maintains constant output voltage (within certain limits) in spite of changing line voltage and/or load current.

Voltage Source. Circuit or device that supplies voltage to a load.

Voltaic Cell. Primary cell having two unlike electrodes immersed in a solution that chemically interacts to produce a voltage.

Volt-ampere. Unit of apparent power in an AC circuit containing capacitive or inductive reactance. Apparent power is the product of source voltage and current.

Voltmeter. Instrument used to measure difference in potential between two points.

Volume. Magnitude or power level of audio frequency. Measured in volume units (VU).

VPN. Virtual Private Network. VPNs use the Internet as the transport backbone to securely connect a business with remote offices, partners and employees from virtually anywhere. VPNs are inexpensive to install and operate, and analysts say that most workers who use them believe their productivity increases when away from the office. In wireless, VPNs are used as an added layer of security — especially for corporations using Wi-Fi or 802.11b wireless networks.

VSAT. *Very Small Aperture Terminal,* referring to satellite earth stations with small antenna diameters, often intended for transportable applications.

W

Wafer. A thin sheet of semiconductor material made by mechanically sawing it from a single-crystal or multicrystal ingot or casting.

WAN. Wide Area Network, typically an inter-city network.

Watt. 1. Unit of electrical power required to do work at the rate of one joule per second. One watt of power is expended when one ampere of direct current flows through a resistance of one ohm. In an AC circuit, true power is the product of effective volts and effective amperes, multiplied by the power factor.

2. The unit of electric power, or amount of work (J), done in a unit of time. One ampere of current flowing at a potential of one volt produces one watt of power.

Wattage Rating. Maximum power a device can safely handle continuously.

Watt-hour. Unit of electrical work, equal to a power of one watt being absorbed for one hour.

Wattmeter. Instrument used to measure electric power in watts.

Wave. Electric, electromagnetic, acoustic, mechanical or other form whose physical activity rises and falls or advances and retreats periodically as it travels through some medium.

Waveform. The shape of the curve graphically representing the change in the ac signal voltage and current amplitude, with respect to time.

Waveguide. Rectangular or circular pipe used to guide electromagnetic waves at microfrequencies.

Wavelength (l). Distance between two points of corresponding phase and is equal to waveform velocity divided by frequency.

Wavelet image coding. Unlike the broadband DCT used for video compression for applications like videoconferencing and DBS in algorithms like H.263 and MPEG2, wavelet coding is generally narrow band, with the coding and filtering tailored to a specific application. Wavelet coding is especially efficient and powerful when applied to phenomena with discontinuous or spiky data. Some wavelet applications include compression of well-defined data sets such as signatures and fingerprints. At the other extreme, wavelet coding may be used for machine vision and more complex applications.

In addition to image processing, the wavelet mathematical functions are widely used for digital signal processing for diverse applications ranging from astronomy to seismology.

W-CDMA (Wideband Code Division Multiple Access). The third generation standard offered to the International Telecommunication Union by GSM proponents.

WCS (Wireless Communications Services). Frequencies in the 2.3 GHz band designated for general fixed wireless use.

Weber (Wb). Unit of magnetic flux. One weber is the amount of flux that when linked with a single turn of wire for an interval of one second will induce an electromotive force of one volt.

Wet Cell. Secondary cell using a liquid as an electrolyte.

Wetting. Term used in soldering to describe the condition that occurs when the metals being soldered are hot enough to melt the solder so it flows over the surface.

Wheatstone Bridge. Four arm bridge circuit used to measure resistance, inductance or capacitance.

Wideband Amplifier. Also called "broadband amplifier." Amplifier with a flat response over a wide range of frequencies.

Wien-bridge Oscillator. Oscillator that uses an RC low-pass filter and an RC high-pass filter to set the frequncy of oscillations.

Wi-Fi. Wi-Fi, also known as 802.11b, is a leading wireless networking standard and operates in the unlicensed spectrum at 2.4 GHz, which is the same frequency band used by cordless phones, microwave ovens and Bluetooth. It uses a direct sequence spread spectrum modulation scheme.

Wi-Fi is capable of transmitting data normally at distances up to about 300 feet at a data rate of 11 megabits per second. Because the spectrum is shared with other users, the rates fall as more users log on. Security is provided by the Wired Equivalent Privacy (WEP) specification, which is relatively easy to break

Wi-Fi5. Wi-Fi5 refers to the 802.11a wireless networking standard, a technology that operates in the unlicensed 5 GHz band and can deliver data wirelessly at speeds up to 54 Mbps. Wi-Fi5 uses the same medium access controller as Wi-Fi but a different physical layer so it is not compatible.

WIN (Wireless Intelligent Network). The architecture of the wireless switched network that allows carriers to provide enhanced and customized services for mobile telephones.

Winding. One or more turns of a conductor wound in the form of a coil.

Window. A wide band gap material chosen for its transparency to light. Generally used as the top layer of a photovoltaic device, the window allows almost all of the light to reach the semiconductor layers beneath.

Windows CE. The Microsoft operating system developed for handheld computing devices. Variations of Windows CE also may eventually become a platform for applications used by communications devices such as wireless phones or two-way pagers.

Winston Concentrator. A trough-type parabolic collector with one-axis tracking, developed by Roland Winston.

Wire. Single solid or stranded group of conductors having a low resistance to current flow. Used to make connections between circuits or points in a circuit.

Wire Gauge. American wire gauge (AWG) is a system of numerical designations of wire diameters.

Wireless. 1. Term describing radio communication that requires no wired between two communicating points. 2. Using the radio-frequency spectrum for transmitting and receiving voice, data and video signals for communications.

Wireless Application Protocol (WAP). A protocol designed for advanced wireless devices allowing the easy transmission of data signals, particularly Internet content, to micro-browsers built into the device's software.

Wireless Internet. An RF-based service that provides access Internet e-mail and/or the World Wide Web.

Wireless IP. The packet data protocol standard for sending wireless data over the Internet.

Wireless IT (Wireless Information Technology). The monitoring, manipulating and troubleshooting of computer equipment through a wireless network.

Wireless LAN (Local Area Network). Local area network using wireless transmissions, such as radio or infrared instead of phone lines or fiber-optic cable to connect data devices.

Wireless PBX. Equipment that allows employees or customers within a building or limited area to use wireless handsets connected to an office's private branch exchange system.

Wirewound Resistor. Resistor in which the resistive element is a length of high resistance wire or ribbon usually nichrome wound onto an insulating form.

Wire Wrapping. Method of making a connection by wrapping wire around a rectangular pin.

WLL (Wireless Local Loop). A fixed service that competes with or substitutes for local wireline phone service.

Woofer. Large loudspeaker designed primarily to reproduce low frequency audio signals.

Work. Work is done any time energy is transformed from one type to another. The amount of work done is dependent on the amount of energy transformed.

Work Function. The energy difference between the Fermi level and vacuum zero. The minimum amount of energy it takes to remove an electron from a substance into the vacuum.

World Administrative Radio Conference (WARC). Biennual meetings of International Telecommunication Union member nations to discuss and resolve global spectrum allocation issues.

WPDA (Wireless Partnership for Donor Awareness). The industry's effort to raise organ and tissue donor awareness.

WTO (World Trade Organization). An intergovernmental organization set up in 1995 to oversee the rules of international trade, thus helping smooth the flow of trade, resolve disputes and organize trade negotiations. The Geneva-based group, created as successor to the General Agreement on Tariffs and Trade, in 1997 negotiated the agreement to open trade and investment in basic telecommunications and information technology products.

X

X. Symbol for reactance.

X axis. Horizontal axis.

X.25. A specification from the Consultative Committee on International Telephone and Telegraph on layered protocols connecting computer terminals to a public, packet-switched network.

xDSL. 1. Designation for digital subscriber line technology enabling simultaneous two-way transmission of voice and high-speed data over ordinary copper phone lines.

2. An unspecified type of Digital Subscriber Line. May be one of ADSL, HDSL, IDSL, MSDSL, RADSL, SDSL or VDSL (which see).

XLR. A circular three-conductor shielded connector commonly used in professional applications for connection of low-level balanced audio signals to and from equipment.

XML. *eXtensible Markup Language,* descends from SGML and is a standard for the exchange of structured and networked data on the Web. XML documents can define their own tags, providing outstanding flexibility.

Y. Symbol for admittance.

Y axis. Vertical axis.

Z

Z axis. Axis perpendicular to both X and Y axes.'

Zener Diode. Semiconductor diode in which reverse breakdown voltage current causes the diode to develop a constant voltage. Used as a clamp for voltage regulation.

Zeroing. Calibrating a meter so that it shows a value of zero when zero is being measured.

Zone Refining. Method of purifying solid rods by means of melting narrow zones through the rods. These zones are slowly moved from one end of the rod to the other, sweeping out the impurities.

Zulu Time. Synonymous with Greenwich Meridian Time, a time designation used in satellite systems.